# BODY STORY

## DR DAVID WILLIAMS

| CLASS | | |
|---|---|---|
| NO. | | |
| AUTHOR | SUBJECT | DATE |
| | | |
| | | |

# BODY STORY

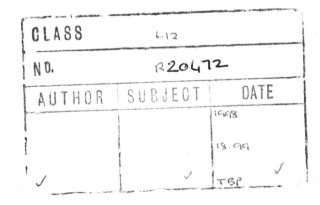

# DR DAVID WILLIAMS

CHANNEL 4 BOOKS

MACMILLAN

First published in 1998 by Channel 4 Books, an imprint of Macmillan Publishers Ltd, 25 Eccleston Place, London SW1W 9NF and Basingstoke.

Associated companies throughout the world.

ISBN 0 7522 2429 8

9 8 7 6 5 4 3 2 1

A CIP catalogue record for this book is available from the British Library.

Commissioning Editor: Charlie Carman
Editors: Christine King and Emma Tait
Design by DW Design
Photography by Pete Jones © Pete Jones
Computer images by 4:2:2 Bristol © Wall to Wall Television Limited
Colour reproduction by Speedscan Limited
Printed in Italy by New Interlitho

This book accompanies the television series *Body Story* made by Wall to Wall Television for Channel 4, the Discovery Channel and ITEL.
Executive producer: Alex Graham
Produced and Directed by Leanne Klein

# Contents

Acknowledgements 6

Preface 7

Introduction 8

THE TAKEOVER 10

CRASH REPAIRS 32

BASIC INSTINCT 56

THE COLD WAR 86

UNDER PRESSURE 110

SHUT DOWN 144

Epilogue 170

Glossary 172

Index 174

Bibliography 176

To Julia, Caitlin, Flora and the other little miracle on the way

# ACKNOWLEDGEMENTS

I would like to thank the production team at Wall to Wall Television for giving me the opportunity to work on such a stimulating and enjoyable project as *Body Story*. Leanne Klein, Dan Gluckman and especially Sajjad Bhatti whose research formed the basis of the book and who were all trememdously helpful and supportive with their opinions. David Burns has done a wonderful job of selecting the images.Others, including Michael Marber and Donald Peebles gave constructive criticism on specific chapters, while my dear friend (and brother-in-law) Andrew Maynard gave invaluable criticism on the whole book.

My special thanks however, go to Christine King with whom I have spent many hours restructuring and rebuilding the contents. Without her this would not have been the book that it is.

Finally, my thanks go to my wife Julia and daughters Caitlin and Flora, whose companionship I have greatly missed while writing this book.

Wall to Wall Television would also like to thank all the consultants involved in the making of *Body Story*: Prof. Christopher Bulstrode, Dr Nicolas Coni, Prof. Anthony Dickenson, Dr Theodora Duka, Dr Duncan Dymond, Dr Ashley Grossman, Dr Eric Jauniaux, Prof. Tom Kirkwood and Prof. John Oxford.

# Preface

When Wall to Wall Television asked me if I'd be interested in making *Body Story* I said 'yes' straightaway. Although I'm not a biologist I could see how much potential there was in the project. Our bodies are made up of hundreds of billions of cells but most of us know little about what goes on beneath our skin. This series was going to try to explain what happens.

Although the series was to be explanatory, we wanted to capture a sense of wonderment, to show events on a scale so tiny that not even the most powerful microscope can see them. *Body Story* was to become the first television series to use 3D computer animation to recreate the moment by moment workings of the human body.

There was one more twist we wanted to give the series. We wanted to tell stories of people like you and me. We decided that each film was to be about a single character going through a familiar biological event. It was to be a series of biological dramas which took you right inside the bodies of its characters.

*Body Story* took just over a year to make, with everyone working flat out. It was the hardest job I've ever done. But I hope the series achieves what it set out to do: to capture the wonder and complexity of the human body, and to show it as it's never been seen before.

Leanne Klein, producer and director of the television series, *Body Story*.

# Introduction

When I first met the production team working on *Body Story* I was immediately impressed not only by their energy and enthusiasm, but also by their whole approach to the project. They were asking apparently simple questions about the human body for which there were no easy answers. We may all – medics included – take our bodies for granted, but many of the fundamental processes that keep us going are by no means fully understood. The team was committed to communicating these complexities in an admirably accessible way, and I welcomed the opportunity to join them as the medical consultant. Their focus on six vital life events was challenging and stimulating. I was particularly impressed by the added dimension given by the computer graphics – they were a revelation to me, illustrating processes that could not be filmed within a living human body.

This book grew out of the series. Each chapter follows the storylines used in each television episode, putting a human face on a medical condition. Here, I have been able to develop the themes covered by the series and include more of the latest medical advances, as well as giving a historical perspective on many different aspects of disease. Like the series, this book does not aim to cover the whole human condition. We look at the beginning of life and its inevitable end and, along the way, the potent relationship between brain and body. We also consider the body's response to three common setbacks: accident, infection and disease.

We first follow a woman from the instant she conceives her baby. From that moment on, mother and child are locked together in the closest possible relationship. We see how the new life makes huge demands on the mother, and how her body must adapt to allow them both to survive. Growing children have a lot to learn on the way to independence. They will make mistakes – accidents will happen. We next look at a common hazard of childhood: a broken bone. From fracture to healing, we follow one of the most remarkable features of the human body – its capacity to repair itself.

Adult life brings further hazards and new temptations. We look at a young man's basic instincts for survival, sex and food. We explore his brain as it controls his body during a night of passion fuelled by alcohol. Accidents and drunkenness may be preventable, but there isn't much anybody can do to avoid infections from viruses and bacteria. We go on to examine the human immune system and its sophisticated battery of defence weapons.

We next look at the most common cause of death in the western world: heart disease. To some extent this is a preventable condition, but we follow one man who has made little effort to avoid his fate. We watch him having a heart attack. Whatever the cause of death, it can strike at any age – though more people are living longer than ever before. Finally we examine the effects of ageing and, in the last hours of one old man, witness the end of the body's story.

Birth and death are the bookends to this story, but each chapter covers some of the most important and familiar biological events of our lives. I hope this book will give the reader a greater insight into the remarkable workings of the human body.

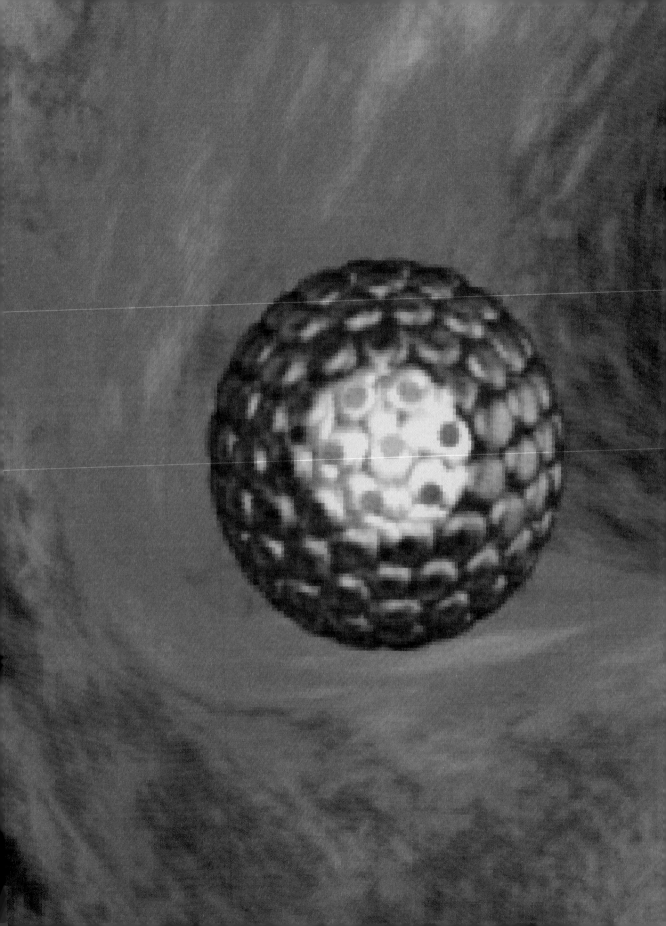

# 1

# THE

# TAKEOVER

A new life begins when a sperm fertilises an egg. For the next nine months the mother's body and mind are in the grip of a life that cannot be sustained without her. This new life is ruthless in its demands on the mother. If these demands are not met, then its very survival is threatened.

efore this uneasy – if rewarding – relationship can begin, the sperm must undergo a lengthy, hazardous journey. It swims a distance 100,000 times its own length – equivalent to an adult male swimming more than 100 miles. Less than one in a million sperm ever complete this journey. Most of them die in the acidic juices of the female vagina, but those that are sucked into the cervix of the uterus (womb) are close to their goal. Propelled by their tails and aided by the muscular contractions of the uterus, sperm make rapid progress to the place of fertilisation – a fallopian tube. One of these tubes passes from each ovary to the uterus.

Amazingly, the ovaries of a five-month-old female foetus contain 7 million primitive eggs – follicles. At birth, this number has fallen to between one and two million and, by the time of puberty, most of these follicles have dissolved, leaving 250,000 available for the reproductive phase of life. At this time, when the hormonal environment changes radically, up to 1000 follicles develop each month, but usually only one develops into a mature egg and is released from an ovary: this is ovulation. In total, each ovary releases about 200 eggs in its reproductive lifetime. (Surprisingly, the ovaries don't take it in turns to release an egg. It is the ovary containing the most mature egg that ovulates next.) Ovulation, unlike male ejaculation, is a slow sticky ooze, and not associated with orgasm. Indeed, it can be a little uncomfortable, causing pain during the middle of a menstrual cycle.

**Previous page:** Six days after fertilisation the embryo is a ball of about 100 cells, ready to embed into the wall of the mother's womb.

**Above:** An ideal cover up for contemplating one's navel. The baby's kicks and punches can be felt after 20 weeks of pregnancy.

## The menopause and HRT

When a woman reaches about fifty years of age, the ovaries stop producing female hormones such as oestrogen and the menstrual cycle stops; she has reached the menopause. In order to prevent the consequences of low blood levels of oestrogen, such as thinning of the bones (osteoporosis) and coronary heart disease (see Chapter 5), some older women take hormone replacement therapy (HRT).

In these women, the endometrium grows as if during a normal menstrual cycle and must be induced to shed artificially. This is done by changing the combination of hormones within the HRT pill at the end of each monthly cycle. If this is not done, the endometrium builds up, putting the post-menopausal woman who still has a womb at increased risk of cancer of the endometrium.

## MONTH BY MONTH

It is interesting to speculate why female reproductive physiology has been designed in such a way that an egg is not released during each orgasm (a belief widely held in medieval times). This strategy would undoubtedly increase the rate of conception and the success of the species. Furthermore, if all eggs available in the infant ovary were still available during fertile years, there would be no shortage of eggs to release. Yet a far more conservative approach has been adopted, limiting the possibility of fertilisation to roughly two and a half days during each month. The simple reason for this cyclical format is that women must carry (literally) the consequences of successful fertilisation for the next nine months. And in order to do this, the woman must prepare her womb to be as comfortable as possible for the developing baby. The menstrual cycle is geared towards creating this environment, in the wall of the uterus.

The first day of the menstrual cycle is, by convention, the first day of a period – menstruation. On average a woman will bleed for five days, losing around an eggcupful of blood. Over a normal reproductive lifetime women shed around 17 litres of menstrual blood; no wonder that some become deficient in the vital ingredient of blood – iron. The bleeding results from the shedding of two-thirds of the mucus membrane (endometrium) that lines the inside of the womb. Once the bleeding has stopped, the one-third of the endometrium left behind acts as a basis for rapid regeneration. It has been calculated that if the endometrium continued to double at the rate it does between day five and day 20 of the menstrual cycle, then after one year the total weight of endometrium produced would approach one tonne! This incredible rate of growth could entail cell division going wrong, leading to malignancies of the endometrium. It is the monthly shedding that usually prevents this happening.

During a normal menstrual cycle, the endometrium is fully restored around day 20. At this time, there is a rich blood supply and an abundance of glands to create an inviting bed for the

### Ectopic pregnancy

In about one per cent of pregnancies, the fertilised egg embeds in an abnormal environment, such as the fallopian tube itself – an ectopic pregnancy. This problem is often the consequence of narrowed tubes, damaged by previous infections in the mother. Tiny sperm can often get through narrowings to fertilise the egg, but the fertilised egg, which starts off life 100 times bigger than a sperm and gets even bigger after fertilisation, is too big to get back down into the uterus. It embeds in the tube, the mother misses a period and is likely to have a positive pregnancy test.

All will appear normal for about four to six weeks, but then the growing embryo becomes short of space and causes pain and bleeding through the vagina. Taken together, these signs and symptoms strongly suggest an ectopic pregnancy. The diagnosis is confirmed by use of an ultrasound scan that can identify the embryo in a fallopian tube. It is treated by clearing out the embryo and repairing any damage to the tube. Often, however, the fallopian tube is damaged beyond repair and must be removed.

In 1994, a medical scandal broke out when a British consultant obstetrician claimed a healthy baby girl was born after he relocated a five-week-old ectopic embryo from a fallopian tube and placed it in the wall of the uterus. Despite publication in the British Journal of Obstetrics and Gynaecology, it was later discovered that the operation never took place. As a consequence the obstetrician was struck off the medical register and the then President of the Royal College of Obstetricians and Gynaecologists who was associated with the incident, resigned. It will probably be many years before this operation can be mastered.

## Infertility and treatments

15 per cent of all couples have less than normal fertility. In cases where problems can be identified, approximately 50 per cent are in the woman, 30 per cent in the man, and the remaining 20 per cent in both partners. However, despite vigorous questioning and investigation, the cause of infertility often remains unknown.

When all efforts to correct the underlying problem in either man or woman have failed, a couple can now resort to in-vitro fertilisation (IVF), a technique pioneered by the gynaecologist Patrick Steptoe (1913–88) and physiologist Robert Edwards (b. 1925). In the late 1960s, they removed an egg from an ovary and successfully fertilised it with sperm outside the body: in the proverbial 'test tube'. Eventually this technique was refined so that they were able to place the embryo back in a mother's uterus. In July 1978 this work was rewarded by the birth of the first ever 'test-tube baby', Louise Brown. It has subsequently proved a useful technique for women with blocked fallopian tubes or men with reduced sperm counts. However, even now, only 15 per cent of attempts at IVF are successful. Recently there have been reports of elderly, post-menopausal women, whose ovaries have stopped producing eggs of their own, becoming pregnant with sperm from their partners and eggs donated by other women. This and the potential use of eggs from aborted foetuses has brought calls for stricter controls on the use of IVF.

Nevertheless, ever more refined IVF techniques are constantly being developed, especially to help men with very low sperm counts. In 1992 a fertility expert successfully inserted a single sperm directly into an egg – a process known as intracytoplasmic sperm injection (ICSI). No need for 400 million sperm with this technique! Indeed, this may be an inherent problem, as we do not know what process of natural selection is being artificially bypassed when one sperm is selected by a technician, rather than the forces of nature. Without any checks on its safety or outcome, this technique has rapidly gained widespread use. Already tens of thousands of children have been born through ICSI, but only now are researchers looking back to investigate the health of these children. Preliminary reports suggest there may be a reduction in the mental ability of one-year-old boys born following ICSI. Whatever is or is not confirmed, it is essential that any new treatments for infertility undergo the same stringent checks that a new drug must endure before finally being released on the market. However, the desperate desire of many subfertile couples to have a child makes it difficult to keep a tight rein on new developments.

fertilised egg. If, however, the egg remains unfertilised, then between days 26–28 the level of hormones released from the ovaries falls dramatically, signalling another round of endometrial shedding.

# GETTING TOGETHER

If fertilisation is going to take place successfully, an egg must have been released from an ovary into the fallopian tube about 24 hours earlier. Released with the egg are hundreds of other cells that provide nourishment while it waits in hope of fertilisation. The egg wafts gently down the fallopian tube to meet the advancing sperm. Whatever chemistry attracts a man and woman remains as mysterious as the forces of attraction between a sperm and an egg. At the time of fertilisation, some sperm swim close to the egg, but pass by without any apparent attraction. Others are guided straight to their target. But even the most focused sperm are incapable of fertilising an egg upon arrival. They have a kind of protective 'head gear' which they must first lose in order to release enzymes that can penetrate the protective barriers of the egg.

Of the 400 million sperm – enough to populate Europe – ejaculated into a woman, only about 400 will complete their journey and reach the site of fertilisation. It seems excessive in the extreme that 400 million sperm are required to allow one to do its job. Yet a man with a sperm count of less than 50 million – close to the population of all England – is considered subfertile. The gross excess of sperm is thought to be necessary to destroy the barrier around the egg. The first sperm to break into the egg triggers a reaction that immediately makes the surrounding barrier impenetrable to other sperm. Hundreds of sperm are suddenly trapped in the wall of the now fertilised egg.

We inherit half our characteristics from our mother and half from our father. Each characteristic is produced by a protein that is generated by a special code held

## Cloning

In 1997 a lamb was born, genetically identical to its 'mother'. The lamb, named Dolly, was a 'clone' of a sheep that had taken no part in a sexual act. There was no genetic contribution from sperm; indeed, the lamb did not have a father. Amazingly, scientists in Scotland had taken an unfertilised egg from a sheep, removed its nucleus and replaced it with the nucleus of a cell from the udder of another sheep. The sheep that donated its nucleus from the udder cell had passed on its entire genetic load to Dolly. A cell taken from almost any part of its body and treated in the same way could have had the same result.

An udder cell can lead to the birth of a perfectly formed sheep because, like almost all cells, it contains a copy of every gene required to generate every protein in the body. However, in any one cell the vast majority of this genetic material lies redundant. Within an udder cell, for example, the active genes might be geared to producing milk, and the genes that produce liver enzymes or brain hormones lie dormant. But if you replace the nucleus of an egg with a nucleus from an udder cell, the chemical environment of the egg stimulates a completely different array of genes. All of a sudden this cell is reprogrammed, firstly to divide, then to transform into an early embryo, before developing all the cell types that make up the adult animal. This is asexual reproduction, where the act of fertilisation by a sperm has been bypassed,

Cloning is an efficient mechanism of asexual reproduction. Cloning of sheep could be used to select animals with particularly advantageous characteristics. However, a single gene mutation in such an asexually reproducing sheep could result in a serious defect that would almost inevitably be passed on to the next generation. Such gene mutations have led to the extinction of asexually reproducing plants, and most notably were part of the cause of the Irish potato famine in 1845. It is for this reason that most animals, including humans, have evolved with two sexes and perpetuate by sexual reproduction. The consequent genetic mix means that deleterious genes do not survive. Nowadays, however, modern medicine is beginning to prove this last statement wrong; children born with inherited diseases such as cystic fibrosis survive until a fertile age and pass on their faulty genes to their children. It is possible, though, that medical science could screen out those faulty genes before conception, leading to the birth of children free from the disorders that they would once have inherited.

Although cloning sheep may be morally acceptable – indeed, a walk through the hills of Wales makes you wonder whether the Welsh have been at it for years – cloning of humans is far more difficult to accept. The spectre of a megalomaniac, cloning himself hundreds of times and passing on the same evil characteristics to his offspring, alarms those of us with a vivid imagination. Tight controls have therefore been put in place to limit experiments on human cloning. However, cloning not of individuals but of groups of cells could be of huge benefit. It is likely that soon we will see the fruits of this research in the treatment of Parkinson's disease.

within a gene. Our genes are carried on our chromosomes held within the nucleus of all cells. However, the nucleus of an egg and a sperm is quite different from the nucleus of any other cell in the body, as it contains only half the number of chromosomes, i.e. only half the genetic material. A fertilised egg contains the full complement of 46 chromosomes (23 matching pairs) that hold a unique mixture of genes inherited equally from mother and father. From this moment on, the genetic features of the new offspring have been established, including whether it is a boy or a girl.

Within the next 30 hours, the fertilised egg splits equally into two separate daughter cells, each containing a nucleus with the full complement of 46 chromosomes. Within 12 hours both these daughter cells divide into two, all new cells also holding the full complement of chromosomes. From here on, the rate of

## Spina bifida and Down's syndrome

One of the most common serious abnormalities in the UK is spina bifida, occurring in about one in 7,000 births. Recent research has now made a link with the condition and a particular vitamin, folic acid. Folic acid is essential for normal cell growth in humans. The explosive growth of cells within the developing embryo means that the pregnant mother must increase her intake of dietary folic acid. Indeed, the foetus will even extract folic acid from a mother who is deficient in folic acid. This is an example of the 'battle' between mother and foetus, where the survival of the foetus – and therefore the continuation of the species – takes priority over the needs of the mother.

Spina bifida is a neural tube defect in which the bones around the spinal cord fail to close over, making the underlying nerves vulnerable to damage. The consequences vary from mild bladder damage to severe neurological damage, including paralysis. Large trials have found that women whose diets were supplemented by folic acid (at least 400 micrograms), from the time of conception until 12 weeks into the pregnancy, are less likely to have babies with spina bifida. How folic acid works is still not understood, but one intriguing suggestion is that it acts by inducing an early miscarriage in those women carrying abnormal babies.

A foetus with spina bifida leaks a substance known as alpha-feta protein (AFP) from the exposed spinal cord, directly into the amniotic fluid that surrounds it. Spina bifida in a foetus can be suspected during pregnancy by high levels of AFP in the mother's blood at 16–18 weeks' gestation. If these blood levels are suspiciously high, a sample of amniotic fluid can be taken to confirm the diagnosis – amniocentesis. This is now a routine procedure but, even in experienced hands, it triggers a miscarriage in roughly one per cent of women. Guided by an ultrasound scanner, a needle is carefully inserted through the mother's abdomen and into the uterus containing amniotic fluid. About 10–20 millilitres of fluid is then removed for analysis. It is not only the level of AFP that can be measured, but the exact chromosome make up of the developing foetus. After sixteen weeks' gestation the foetus, like any adult, starts to shed its skin. Of course, unlike adults, the foetus sheds its skin into the surrounding amniotic fluid, and it is the chromosome make up of these skin cells, rescued from the amniotic fluid, that can be analysed.

The most likely reason for doing this analysis is to diagnose Down's syndrome, an abnormality characterised by a third copy of the twenty-first chromosome. Once the diagnosis has been made (it takes three more weeks for the results to come through) the parents face the difficult decision about whether or not to continue with the pregnancy. For some, especially those with strong religious beliefs, there is no issue. Life is sacred in whatever form and must always be protected; of course this opinion negates any use for amniocentesis. Others take a more pragmatic approach. They count the extra emotional, physical and literal cost of caring for a mentally and often physically handicapped human being for the rest of their lives and consider the price is too high to pay.

The six day old embryo is still only a ball of cells. The outer cells eventually become the placenta which produces enzymes. This allows the embryo to burrow deeply into the wall of the mother's womb. On the left, the embryo is burrowing in, on the right it can be seen embedded.

## Negative *v* positive

The embryo is separated from maternal blood by a barrier only one or two cells thick. If the mother's blood crossed into the circulation of the developing foetus, we would expect all babies to have the same blood group as their mother, but this is not the case. Occasionally, at birth or at the time of a miscarriage, when the placenta comes away from the lining of the womb, blood from the foetus enters the mother's circulation. This is of no consequence if the mother and foetus happen to share the same blood group. However, the foetus may have inherited its father's blood group.

The most striking example of this problem is when the foetus has the rhesus positive blood group and the mother, like 15 per cent of the Caucasian population, is rhesus negative. When this occurs, the mother sets up an immune response to kill off any foetal red cells that have escaped into her circulation. This leads to severe problems for a subsequent foetus who may also be rhesus positive. Having been primed by the earlier pregnancy, the mother's immune system is ready to unleash its full force on the rhesus positive red cells of the next foetus. Antibodies created by the mother, specifically targeted at rhesus positive red cells within the foetus, can cross the placenta to cause a severe and often fatal anaemia. This is known as haemolytic disease of the newborn. It can be prevented by injecting the rhesus negative mother with an antibody (anti-D immunoglobulin), within 72 hours of a delivery or miscarriage. Anti-D immunoglobulin kills any rhesus positive red cells that might have entered the mother's circulation from the foetus, before she can mount an antibody response. Anti-D immunoglobulin is now also routinely given to all rhesus negative mothers at 28–32 weeks' gestation.

cell division starts to speed up and no longer happens synchronously. A ball of cells is formed (under the microscope it looks like a raspberry). It takes about four days for this dividing ball of cells to travel along the fallopian tube and reach the womb.

This rapid cell division goes on for the first week after conception, but the mother is completely unaware she is pregnant. Within six days the embryo is a ball of 100 cells, free-floating in the womb. The mother's womb is thick with blood vessels and glands which grow each month as part of her menstrual cycle. In a week's time this lining will be lost with her next period, taking with it the fragile embryo. To prevent this happening, the embryo must first 'take control' of its mother's body, burrowing deep into the tissue lining of the womb. The outer cells, soon to become the placenta, release chemicals that destroy just enough of the womb lining to let the embryo inside.

## Pre-eclampsia

Pre-eclampsia is a syndrome that has its origins in early pregnancy, but does not manifest itself until late pregnancy. It affects about five per cent of first-time pregnant mothers and is recognised by high blood pressure and protein in the urine. However, these abnormal signs usually go unrecognised by the mother, until picked up by a routine antenatal check. The only definitive treatment for this condition is delivery of the baby. This can be a very difficult decision, especially when involving the delivery of a premature baby whose chances of survival are slim. However, nowadays, babies as young as 22 weeks (five months) gestation have survived a premature delivery. Nevertheless, pre-eclampsia is often missed until the mother is very sick, or has developed full blown 'eclampsia' – recognised by convulsions.

## A POSITIVE TEST

A normal menstrual cycle is triggered by the ovaries when they switch off production of oestrogen and progesterone. But within days of conception, high levels of a hormone called human chorionic gonadotrophin (HCG) released by the placenta have hijacked this monthly event. The ovaries continue to pump out their hormones. As the mother's blood is flooded with HCG, large amounts appear in her urine. It is these high levels of HCG that are detected in the urine by modern pregnancy-test kits. In 1920, two gynaecologists from Berlin found that urine from pregnant women must contain potent reproductive hormones. They diagnosed pregnancy by injecting women's urine into laboratory rats; urine from pregnant women caused the animals to go on heat. Thankfully, we have moved on from requiring a pet rat to confirm pregnancy. Modern pregnancy-test kits are now so sensitive they can allow a woman to know she is pregnant within 24 hours of missing a period.

HCG has other more dramatic and unpleasant effects on the mother's body. It is almost certainly responsible for making her nauseated and triggering the vomiting centre situated in the primitive part of the brain.

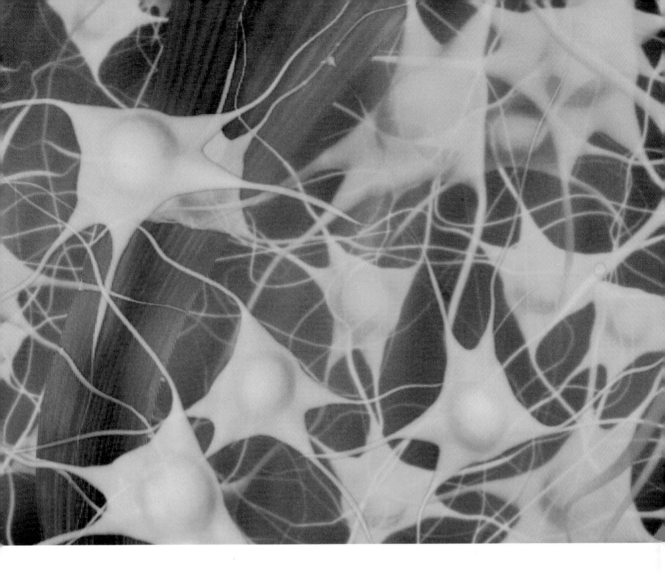

## The first three months

By the twenty-third day of gestation, the embryo is only a few millimetres long, yet it has a beating heart and a head with eyes, ears and a mouth. As most foetal organs develop between the fourth and eighth weeks, this is the period when the foetus is most vulnerable to factors interfering with development. Congenital malformations seen at birth usually originate during this critical period. Some foetal genes have abnormalities that are incompatible with life; this is the usual cause of miscarriage (more than half of all pregnancies end in miscarriage within 12 weeks).

By six weeks, the foetus is pea-sized. The placenta meanwhile has grown even faster and is, relatively, much bigger than the embryo. This ingenious structure, created as a by-product of fertilisation and then unceremoniously discarded nine months later, is more versatile than any other human organ. It combines the key

While still only a few cells in size, the tiny embryo has far-reaching effects on the mother's body. HCG released into her circulation stops her from menstruating. It also affects her brain (shown here) triggering vomiting, a common symptom of early pregnancy.

## Eating for two?

Debate continues about the ideal amount of weight gain during pregnancy. Should a mother eat for two, or carry on as normal? In 1990 the US Institute of Medicine recommended a liberal gain in weight of 11–16 kilogrammes for women of normal body-mass index (see Glossary). These recommendations aimed to prevent undernourishment of the foetus. However, the causes of low birthweight are far more wide-ranging than just undereating by the mother: e.g. heavy smoking, drug abuse, poor antenatal care. Rather than overfeed millions of women, and fuel the western world's epidemic of obesity, it would seem more sensible to address the root causes of the social circumstances that lead to low-birthweight babies.

A more prudent approach is that there's no need to change eating habits (assuming these are sensible to start with) until after the sixth month; by this time the foetus is growing rapidly and a modest increase is warranted. In total, a weight gain of no less than seven kilogrammes and no more than 11.5 kilogrammes is recommended.

functions of a stomach, lungs, liver and brain for the embryo, as well as acting as a gland producing a variety of hormones and enzymes to keep the mother in check.

The healthy embryo influences its mother's body over her whole pregnancy, bringing about changes greater than any she could otherwise experience. Within the first month, her body has begun to change at a significant rate, out of proportion to the current needs of her embryo. The first thing she is aware of is a definite enlargement and tenderness of her breasts. Even though she won't be breastfeeding for another eight months, they are already growing the complex ducts and structures that will produce milk. Ultimately, they may double in size. Under her skin, the changes are even more dramatic. The embryo's placenta has started to produce its own progesterone and oestrogen. Soon it will completely take over this role from the ovaries. Her bloodstream is being flooded with these hormones that make muscles all over her body relax. Even the tiny muscles in the walls of her blood vessels relax, leading to a fall in blood pressure. Consequently less blood gets to her brain, occasionally making her feel dizzy and light-headed, especially when standing up quickly. All these symptoms are more pronounced in a mother who carries more than one growing embryo.

In order to nourish the foetus during its development, the blood supply to the mother's womb increases more than 10-fold. Jets of oxygen-rich blood pass through fine spiral arteries within the wall of the womb into blood lakes, created at the interface between womb and placenta. Oxygen and other nutrients diffuse across the cells at the edges of these lakes and are carried away into the foetal bloodstream. The umbilical vein carries oxygenated blood to the foetus, while the umbilical artery brings back deoxygenated blood to the placenta.

As half of the embryo's genes come from the father, the embryo is half alien to the mother's body. Normally, her body would reject anything it didn't recognise. But the embryo has evolved a remarkable survival strategy, making itself invisible to many of the immune cells which police the womb looking for foreign invaders. The placenta invades the inner lining of the womb; placental cells crawl into the mother's uterine blood vessels and make them much wider, to carry more blood to the developing foetus. Some placental cells break away and get dragged into the mother's bloodstream. Placental cells can travel anywhere around the mother's body – even into her lungs or brain. However, excessive numbers of placental cells within the maternal circulation have the potential to trigger damage to the mother's blood vessels and a condition known as pre-eclampsia (see page 18).

The lowering of maternal blood pressure has another effect: subconsciously the mother's body reacts as if it were losing large quantities of blood from a wound. A survival response is triggered, stimulating the centres in her brain to make her feel thirsty. More than a litre of extra liquid – plasma – is channelled into her dilated blood vessels, making her less prone to faint. However, she could not survive with such dilute blood, so more red blood cells are also manufactured. The factories that produce red blood cells, deep inside the mother's bone marrow, go into overdrive. This was the embryo's objective all along. The new red cells deliver extra oxygen that it needs to grow. They are slightly larger than red blood cells from non-pregnant women, because they absorb more liquid and swell. Their extra size makes them more efficient at carrying oxygen. However, despite this extra blood, its volume doesn't quite fill her

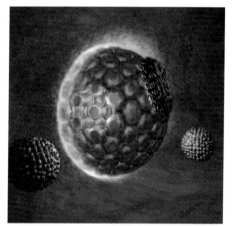

These pictures show the mother's immune cells approaching the embryo. However, they do not recognise the developing embryo as 'foreign', even though half its genes come from the father. This is an amazing adaption as normally anything 'foreign' – such as a heart transplant (see Chapter 5) – would be fiercely rejected unless the recipient's immune system was artificially suppressed by drugs.

dilated veins and arteries until much later in pregnancy. In an attempt to keep her blood pressure normal, her heart must work much harder and beat faster.

Her newly increased blood flow is responsible for a transformation in her senses. The blood vessels in her nose dilate, giving her a hugely heightened sense of smell. This also explains why some women complain of frequent nose bleeds during pregnancy. The mother begins to feel exhausted. The overdose of progesterone in her body is seven times its normal level. It acts like a tranquilliser, slowing down her brain activity and making her sleepy.

Eight weeks after fertilisation, the embryo becomes a foetus. However, this change in name does not correspond to any sudden physical metamorphosis; it marks the stage at which all vital organs have been formed, in a recognisable if primitive state. From now on these organs must simply mature and get bigger. After 12 weeks of pregnancy, all the foetus's organs, including its brain, are beginning to function, yet it is still only six centimetres long.

During the first three months of pregnancy, the mother's body has undergone the most remarkable physiological transformation. Her core body temperature has

## Labour pain and relief

The pain of giving birth is an evolutionary pay-off for humans. No other mammal experiences such agony. We humans walk upright and so we need a rigid, relatively inflexible pelvis to support us. We are also the species with the largest brain, so our babies have the largest heads. The force of the baby coming down the birth canal stimulates a mass of pain receptors, which send impulses shooting up the mother's spinal cord and into her brain. The pain is so extreme that she often becomes confused by its true source and sometimes even her legs seem to hurt.

However, pain during labour is perceived very differently by different women. A complex interaction of physical, psychological and social influences combine to set a unique pain threshold. We have come a long way since Queen Victoria, in the 1850s she was given a general anaesthetic for the birth of one of her children. Nowadays, many different options of pain relief are available. The most accessible is a 50:50 combination of nitrous oxide (laughing gas) and oxygen, frequently referred to as 'gas and air'. The labouring mother can inhale vigorously, achieving a good level of pain relief, without impairing the labour or harming the foetus.

However, the most effective pain relief in labour is undoubtedly provided by an 'epidural block'. This involves a local anaesthetic in the skin over the lower back to allow insertion of a bigger needle into the epidural space. This space contains the nerves that carry pain impulses. Once the needle is in, a very fine flexible tube is inserted through the needle into the epidural space and then the needle is removed. An anaesthetic is then instilled through the fine tube to numb the pain fibres. The pain relief lasts for one to two hours, after which time it may need to be 'topped up' with more anaesthetic. Although usually very effective, the loss of sensation means that women with an epidural are unaware when they are having a contraction. During the second stage of labour, she must therefore rely on the midwife to tell her when contractions occur. If necessary, even caesarean sections can be performed under epidural block. An intravenous line is necessary for women who have an epidural, as often blood pressure falls when the nerves are anaesthetised. This 'side effect' can be advantageous in women who have high blood pressure.

risen by 1°C, her hands are 5°C hotter and her heart is pumping an extra 4,000 litres of blood each day. Had her body made this huge leap overnight, it would almost certainly have killed her. Mother and foetus have reached a crucial stage. The risk of miscarriage is now minimal. Her body has been forced to accept its guest and is fully equipped for the events to come.

For a while mother and baby have reached a comfortable stand-off. The levels of HCG in her bloodstream have fallen, as the placenta produces enough oestrogen and progesterone of its own. The mother no longer feels nauseated or exhausted, but the foetus is still calling the shots. At three months, it weighs only about 15 grams; over the next six months it must put on at least another 3,000. Every new cell in its body must be built using nutrients which only the mother can provide. The foetus demands large quantities of food from the mother's bloodstream; as a result her blood sugar levels drop much faster after a meal than they used to and her appetite has started to soar. As it grows larger, the foetus demands more and more food from the mother's blood.

## The second half

At around four and a half months a new strategy comes into play; this time the 'weapon', also a hormone, is human placental lactogen (HPL). Whereas in the first half of pregnancy the mother is sensitive to insulin and lays down fat stores, in the second half HPL causes her body to become resistant to insulin. This means that the blood stays sugary for longer after she eats, allowing the foetus to gorge itself on sugar at leisure. The mother's body fights back, pumping out more and more insulin, in the only way it can to bring her blood sugar level back to normal. Some women become particularly resistant to insulin and develop dangerously high blood sugar levels – so-called gestational diabetes mellitus. This problem, which disappears after the baby is born, affects about one in 20 pregnant women and requires the mother to

## Alternative deliveries

Over the last 25 years in the UK, the number of home deliveries has declined from about 50 per cent to one per cent. About one in four women who plan a home birth have to be transferred to hospital for failure of labour to progress or distress to the foetus. For these reasons, only women who are at low risk of complications should plan a home delivery. The main concerns about home delivery are the unexpected complications of labour. These include severe haemorrhage, and poor health of the baby following distress during delivery. The midwife and other primary carers must therefore be expert at dealing with obstetric and neonatal emergencies.

In 1992, a UK government health committee recommended that all obstetric hospitals should provide women with the option of a birthing pool. Obstetric hospitals complied, but to date there is still no good information on the benefits and dangers of this method of childbirth. Furthermore, there is nothing natural about giving birth under water. No women of anthropologically primitive groups have ever given birth under water and neither do primates. The only mammals that do, such as whales and dolphins, have little choice. But even whales give birth close to the surface so that the newborn calf can quickly breathe air. The only mammal that chooses to give birth under water is the giant hippopotamus, and here the aim is to soften the landing for its weighty offspring! It has been suggested that because a foetus is within water in the womb, it is more natural for it to be born under water. This is not true, as the constituents of amniotic fluid within the womb are very different from the tap water in a bath. As soon as a baby is delivered, it loses any oxygenation from the placenta (squeezed upon by the contracting uterus) and must gasp for air to survive. A lungful of water leads to drowning or inhalation of maternal faeces – a frequent accompaniment to childbirth and a recipe for infection. Women may find it comforting to climb into a warm bath at the onset of labour, but would be wise to leave it for delivery.

avoid sugar in the diet and sometimes take extra insulin by injection. It is more common in women who were very overweight before pregnancy, and who have a tendency to develop diabetes in later life.

The fat stores have a purpose; they will be used to create breast milk for the baby once it is born. A woman of average weight will put on at least four to five kilogrammes of fat over the course of her pregnancy, even though her metabolic rate has increased by 15 per cent and despite the fact that so much of the food she eats goes straight to her foetus. In total, over nine months, the average western mother will have gained about 12.5 kilogrammes in weight – most of which is water.

Here we can see huge levels of oestrogen and progesterone circulating in the mother's bloodstream during pregnancy. Oestrogen relaxes the tiny muscles in the walls of the blood vessels and progesterone relaxes the ligaments in her pelvis and spine – occasionally causing backache.

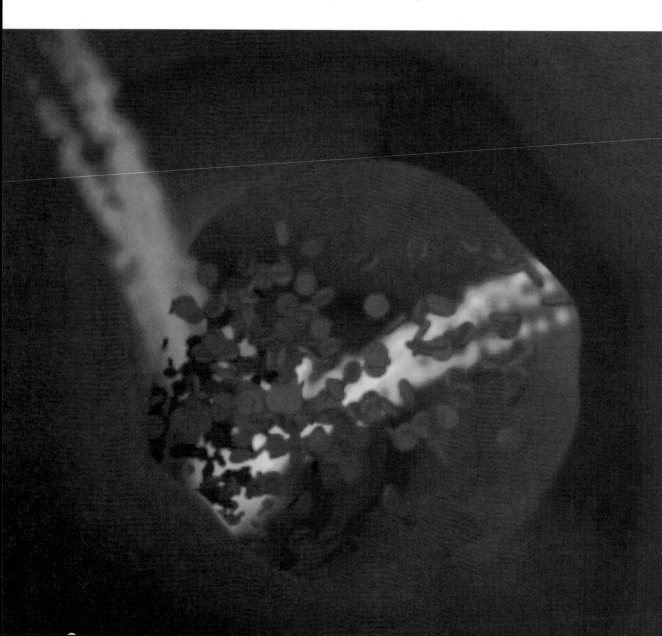

## Does mother always know best?

A recent notorious case highlighted the difficulties in deciding whether a mother has the right to decide the fate of her unborn baby. A 30 year-old single mother was eight months pregnant when she first visited her general practitioner. After examining her, the GP diagnosed pre-eclampsia and recommended that she go to hospital immediately for the baby to be delivered. The mother, a healthcare worker, who objected to invasive procedures in pregnancy and had planned a home birth, refused. The GP called a social worker who arranged for her compulsory admission to a psychiatric hospital. From there she was transferred to the obstetric department of a major teaching hospital where her baby daughter was delivered by caesarean section. Permission to perform the caesarean section was obtained from a judge, despite the patient refusing her written consent on three occasions.

An appeal court has subsequently ruled that a woman who is mentally competent to refuse treatment has a right to do so, even if the result is certain death for herself or the foetus.

Most of us acknowledge that it is the right of every sane adult to refuse medical treatment or advice. The difficulty with this case arises out of the involvement of an innocent party, the unborn child. Once born, the child becomes the legal responsibility of its guardian, the mother. If, for example, the mother has peculiar religious beliefs and forces them upon her child, so endangering its health, the child can be made a ward of court and allowed treatment. Until a baby is born it has no rights of its own. Therefore the mother who refuses treatment herself, denies the obstretician permission to reach the unborn child – however disastrous the consequences.

### Feeling good

Half-way through her pregnancy, the mother starts to enjoy some of the benefits of her hormone overload. Over the nine months she will produce more oestrogen than a non-pregnant woman could make in a hundred years. Oestrogen makes her feel happier, by sensitising her brain to the mood-enhancing chemical, serotonin. (It is the sudden fall in oestrogens after delivery that makes women vulnerable to post-natal depression.) The increased blood supply to her skin may give her a 'glow', while her lips are a little redder than normal and her hair is lustrous. Each hair grows thicker and faster during pregnancy. This may be an evolutionary adaptation, making her more attractive to her partner – encouraging him to stay until the end.

After three months of pregnancy the grapefruit-sized womb expands rapidly upwards out of her pelvis. Despite this 'big bang', no new muscle cells have grown within the mother's womb. Each muscle cell has stretched and will continue to do so until it is five times its normal length. The displacement of organs within her body is astounding. Her entire internal anatomy has been reorganised to accommodate her foetus. Even her heart, already enlarged with an increased workload, has been shunted sideways. Her lungs have also been pushed upwards in the move, yet they too are having to work harder than ever to satisfy the demands of her foetus. Waste products from the foetus, which are pumped through the umbilical cord into her bloodstream, include high levels of carbon

As the lungs are pushed upwards and the heart made to work harder, even light exercise can seem like hard work. But gentle swimming which takes the weight off the legs is a healthy activity.

dioxide, a gas which could cause her serious damage if her breathing rate did not increase to get rid of it. As a consequence she must inhale and exhale 20 per cent more air than normal. Now she becomes breathless with the slightest exertion. Other waste products are excreted through the kidneys, which filter 50 per cent more blood and keep it in a cleaner state than usual. However, the greatest casualty of the reorganisation is the mother's bladder. From early on, her enlarging womb has been pressing down on it, creating a sense of urgency whether or not she actually needs to pass urine.

## The greedy foetus

In the final three months of pregnancy, the foetus grows faster and faster. By seven months it weighs up to 1.5 kilogrammes and will continue to put on about 200 grams a week until it's born. Now its tiny bones are beginning to harden, but at the mother's expense. The foetus is taking calcium from its mother's bloodstream to build its bones. In partnership with her kidneys, the placenta produces extra vitamin D, which makes the mother absorb more calcium than usual from the food she eats. But the quantity of calcium in her food is not enough. The mother's bones have started to give up their own calcium stores to the foetus. A natural process of replacement and renewal has been hijacked. The cells that dissolve calcium in her bones go into overdrive and release the mineral into her bloodstream. The cells which normally replace calcium in the bones temporarily lie dormant. As a result, there is five per cent less calcium in the mother's bones than normal, a level that declines further if the mother breastfeeds.

For months the foetus has also been taking proteins from its mother's blood to build its cells, and the side-effects are beginning to show. One of the normal jobs of protein in the blood is to prevent plasma leaking through the walls of blood vessels into the surrounding cells. As the foetus uses up this blood protein, the mother's tissues have begun to swell with liquid. Swelling at the feet and ankles is common; what's more surprising perhaps is that it can also occur at the wrists. Here an important nerve – the median nerve – that supplies some of the muscles of the hand travels among other tissues through a fibrous band called the carpal tunnel. Swelling

This shows oxytocin – the hormone that is released from the baby's and mother's brain simultaneously – in the bloodstream. The hormone triggers labour, causing the muscle cells of the womb to contract, and push the baby through the pelvis. Synthetic oxytocin can be given by an infusion to induce labour in the mothers for whom it doesn't start spontaneously.

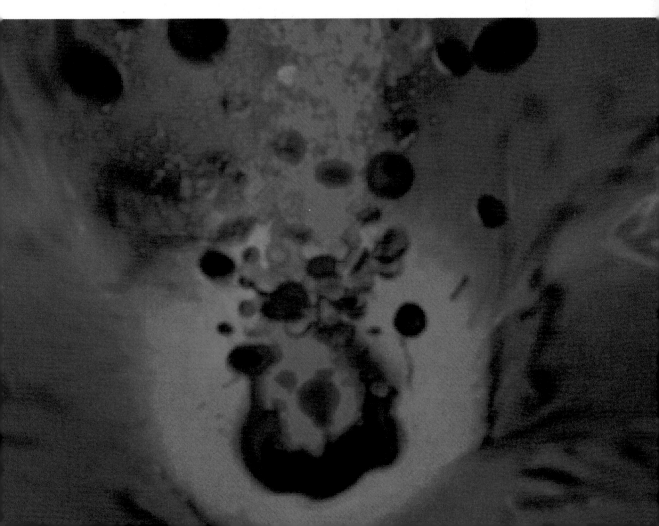

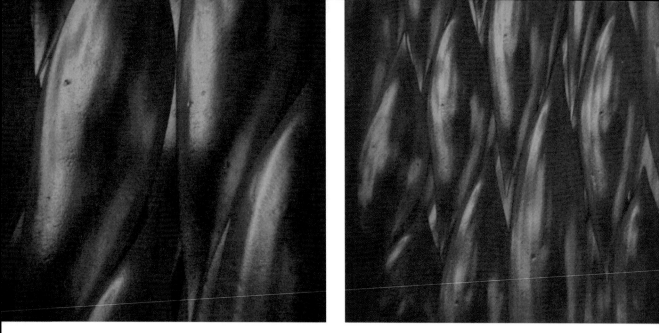

at the wrists causes compression of the median nerve. As a result, towards the end of pregnancy, some fingers and the thumb become numb and often painful, especially at night. Carpal tunnel syndrome almost always disappears as soon as the baby is delivered. In the meantime relief can be obtained by keeping the wrists straight using splints, or by an operation that releases the pressure by cutting the fibrous band.

By eight and a half months, the foetus is 40 centimetres long. For the last three months it has been hearing its mother's voice (and her taste in music). Now it is even sensitive to changes in light from the outside world. The nervous system is well developed and the foetus is flexing its muscles – literally.

Towards the end of pregnancy the mother becomes aware of a tightening sensation known as Braxton Hicks contractions. These mild irregular contractions are like a work-out for the womb muscles, keeping them toned and fit in preparation for the exertions that lie ahead. They've been happening since she was nine weeks pregnant, but only now are they strong enough for the mother to feel. Her body is adapting to make her birth as easy as possible. The ligaments which hold her bones together are softening. Within the pelvis this is a crucial preparation for letting the baby out. It also makes her spine more flexible, helping her adjust her posture. However, this spinal flexibility can often be the source of aching in the lower back. As her centre of gravity has changed, she's started to walk with a characteristic waddle.

The foetus has reached a critical point in its growth. Its head is now nine centimetres in diameter; if it grows any more it will not be able to escape through

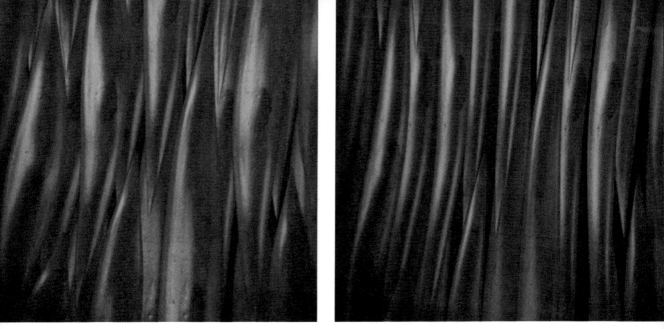

the mother's pelvis, which usually has an inner diameter of only 10 centimetres. So far, mother and foetus have mostly been 'battling' for their respective needs; now they start to work in unison, and must continue to do so if both are to survive.

## LABOUR OF LOVE

The foetus sends off a chemical signal which sensitises the mother's womb to the hormone oxytocin. Oxytocin is released from the brains of both the mother and foetus, triggering contractions. The cervix at the neck of the womb is being pulled and forced to open by the contractions. It will dilate by around 1.5 centimetres per hour for the next six hours. The contractions become increasingly frequent and regular. At the start of labour, they may occur every 15 minutes and last for 30 seconds but, as labour progresses, the interval between contractions shortens and their duration lengthens. With each one, more oxytocin is released, triggering ever more forceful contractions. Just prior to delivery, contractions occur as often as every 90 seconds and last for 30 to 60 seconds. Once it has started, labour cannot stop; it only becomes more intense until the baby is born. Occasionally labour is triggered before the baby is mature enough to survive outside the womb. Drugs can be used to stop early labour, but they are often overwhelmed by the natural forces that impel a premature delivery.

The mother is not the only one in distress. Every contraction causes a massive build up of pressure in the womb, trapping blood within the placenta and umbilical cord. As the seconds tick away, the baby quickly uses up the limited oxygen trapped in its own circulation. The pain of a contraction for the mother is cramp as her

The pain of labour is caused by the contracting muscle cells of the uterus being starved of oxygen.

womb muscles are starved of oxygen. For the baby, its oxygen supply has been cut off until the contraction eases.

The baby's head is forcing its way down the birth canal. It is always a tight squeeze but, when the going gets tough, mother and baby use various tricks. Only during human births must the baby turn 90 degrees as it travels down the birth canal. It ensures that as the baby's head progresses through the pelvis, its widest part passes through the widest part of the pelvis. Furthermore, it is not only the ligaments holding together the mother's pelvis that stretch, but the baby's head also changes shape. The powerful contractions of the uterus cause the flat bones of the baby's skull to slide over one another and the scalp to be moulded into the shape of the path it takes through the mother's pelvis. After what may be many hours of labour the mother's cervix becomes fully dilated – this marks the end of the first stage of labour.

Once the cervix is fully dilated, the much shorter second stage of labour begins – usually lasting not more than one hour. The top of the baby's head normally comes into view first, followed by the forehead and face. Once in the outside environment, the head rotates through another 90 degrees, allowing the shoulders to do the same within the pelvis.

During the stress of labour a baby's heart rate, changing in response to the contractions, can be monitored directly. This will tell the obstetrician and midwife whether the baby is in distress and therefore whether it needs urgent delivery. If this is the case, the options are to perform an emergency caesarean section or assist the mother with a vaginal delivery using forceps or a suction mechanism known as ventouse extraction. The latter involves a plunger device placed on the top of the baby's head which adheres by suction. The obstetrician can then pull out the baby

without the awkward process of fixing forceps either side of the baby's head. The only disadvantage of this technique is that it brings about the most pronounced swelling of the scalp, necessitating a careful angle on photographs of 'the most beautiful baby ever born'!

In England the number of babies delivered by caesarean section had risen from about two per cent in 1955 to 15.5 per cent in 1995. This mainly reflects an increase in the number of emergency rather than planned operations. None the less, more than 70 per cent of all deliveries are performed by midwives and 10 per cent of women leave hospital on the day they give birth, while another 27 per cent leave hospital the next day.

If all goes well, after 280 days or 40 weeks, a baby has been born. As the baby breathes air with its lungs for the very first time it is no longer dependent on its mother's body for survival. The closest of all relationships ends, but a new one is about to begin.

It doesn't take long before all the pain is forgotten and a new set of emotions come into play.

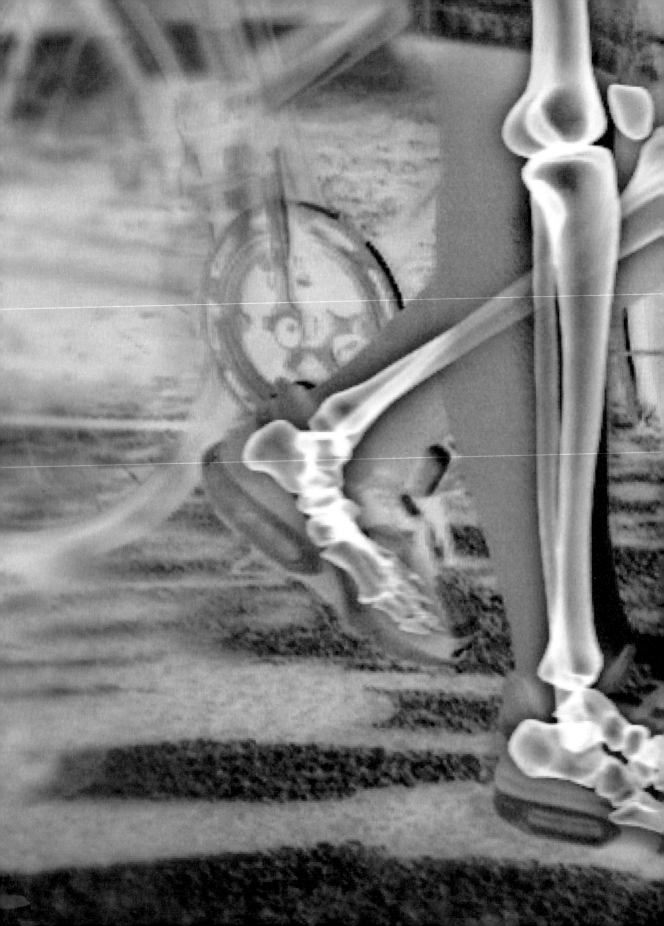

# 2
# CRASH
# REPAIRS

The human skeleton is a triumph of bio-mechanical engineering. Over 200 perfectly shaped bones are linked together by more than 100 movable joints, to give both flexibility and agility combined with strength and support. Every day it is put under tremendous strain. If it were a machine it would break down in months. But the human skeleton is incredibly strong and also has an amazing capacity for self-maintenance and repair.

## The importance of calcium

When calcium is released from bone into the bloodstream, it plays a critical role in the control of nerve impulses to muscle. Too little calcium causes the muscles to go into spasm – a condition called tetany. (This is similar to the muscle spasm caused by the bacterium C. tetani; see page 51.) Too much calcium causes a slowing down of these nerve impulses, leading to paralysis. Inside cells throughout the body, calcium is found at very low levels, but here tiny changes in the level of calcium are of vital importance. Hormones and enzymes are triggered into action by the slightest change in calcium concentration. Calcium is even crucial for the division of cells.

The importance of these functions has prompted clinical scientists to look at ways of manipulating the flow of calcium in and out of cells. Several drugs have now been developed that interfere with the passage of calcium into the muscle cells that line blood vessels. These drugs, known as calcium channel blockers, cause the muscles to relax and consequently high blood pressure falls. They are now in use throughout the world for the treatment of high blood pressure.

**Previous Page:** Bones are linked to each other by ligaments, and to muscles by tendons. The complexity of movement at the hip, knee and ankle joints gives every human a unique walk.

Bones provide a supporting framework for the whole body, as well as protecting vital internal organs and providing points of attachment for muscles so that we can move. They also have another function: they store more than 99 per cent of the body's calcium. Bones must combine strength with sensitivity, responding to a delicate balance of vitamins and hormones in order to keep the blood levels of calcium within a narrow range. There is little margin for error. If blood calcium levels drop too low, then muscles start to twitch and spasm; when heart muscle is involved, this can be fatal.

Bones and teeth are the only tissues that rely heavily on calcium minerals to function normally. The thought of calcium – hard, stony and cold – and the familiar stark image of an old skeleton long after death, stripped of its fleshy covering, give the impression that bones within our living bodies are similarly lifeless. Nothing could be further from the truth. Beneath the surface, bones are teeming with life: full of blood vessels and nerves, they are as alive as our heart or brain. Millions of microscopic lifeforms work tirelessly to make and maintain the intricate structure of our bones.

Bones can resist compression forces twice as well as granite, yet they are just a fifth of the weight of steel. Deep inside, belying its superficial appearance, bone isn't solid. It's made of interconnecting arches like a honeycomb, housing nerves and blood vessels. These arches are built of a composite material – a mixture of tough minerals like calcium and a flexible protein called collagen. Collagen gives a characteristic flexibility to the nose and ears. Without collagen, our bones would shatter like glass; without calcium, they would bend like rubber. Only during this century did human engineers discover a fact that nature has always exploited: hollow structures made of composite materials are the strongest.

The structure of our bones continually adapts to the way we use them. For example, every human has a unique walk, and the leg bones have evolved a unique structure to match. Inside, the arches have grown in a pattern designed to disperse the impact of the way we walk and to protect our delicate joints. If we changed the way we walk, perhaps through injury, these arches would realign themselves.

The arches are constantly renewed: the huge workforce of microscopic organisms ensures that our skeleton is kept in a state of eternal youth. Three types of bone cell are involved – osteoclasts, osteoblasts and osteocytes; these continually communicate with each other, working usually in perfect harmony so that when one has finished its job, another can start.

First, osteoclasts – bone-destroying cells – squirt hydrochloric acid (strong enough to burn through sheet steel) on to the old bone. As the osteoclast dissolves away the tough mineral coating and the strands of collagen beneath, a crew of osteoblasts – bone-building cells – follows in its wake. Triggered into life, these activated osteoblasts change shape, turning from flat, apparently inactive cells into

An osteoclast destroys bone with concentrated hydrochloric acid. This begins the cycle of bone renewal that affects 10 per cent of our bones every year. This means our skeleton is never more than 10 years old.

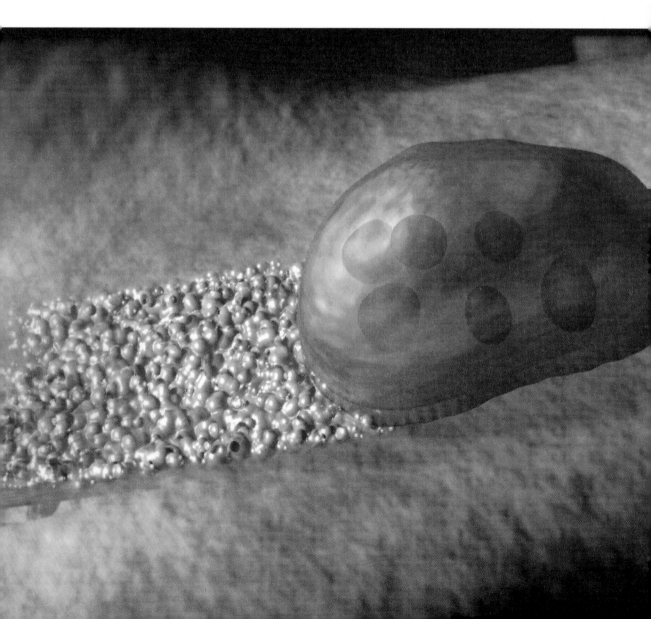

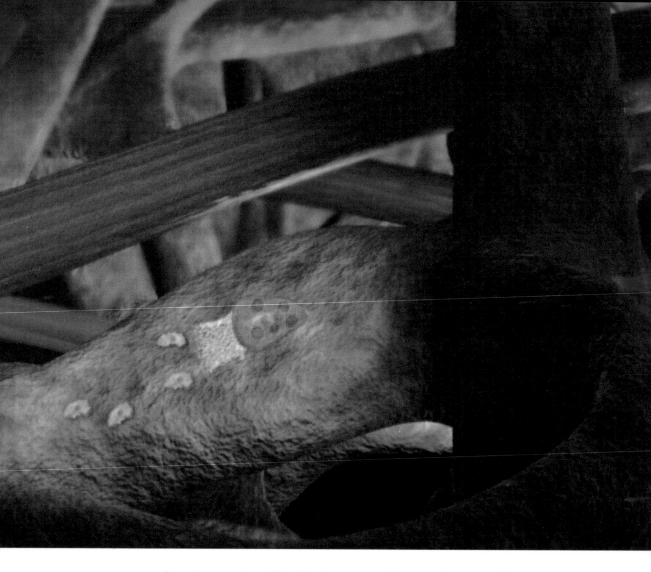

plump bone-makers. They lay down new collagen and coat it with a fresh frost of minerals. When they have finished their work, osteoblasts literally wall themselves in with new bone and become osteocytes. These trapped cells send out numerous filaments like tiny antennae to 'broadcast' from their prisons of bone. Throughout the skeleton thousands of trapped osteocytes detect subtle changes in mechanical forces and then send out signals to communicate with other cells and orchestrate the need for bone repair. They are part of the complex process by which our bones adapt to the stresses and strains that we put them under.

Throughout the body's skeleton, the destroyers and the builders of bone work ceaselessly, to prevent bones from ever getting so old that they crack and crumble. In a young adult, there may be several million sites of bone renewal at any one time. However, at this stage of life the bone mass is constant, so the amount of

## Osteoporosis: causes and remedies

Several factors influence the risk of developing osteoporosis. More than thirty years ago the late Professor Charles Dent, a past professor of medicine at University College, London, remarked, 'Senile osteoporosis is a paediatric disease.' By this, he meant that those who have a greater density of bone in early life can afford to lose more bone in later life; they have a greater reserve of bone to withstand the inevitable loss. New research is proving him right. Osteoporosis runs in families, and the inherited differences in response to vitamin D, a vitamin that controls the amount of calcium in our bodies (see page 40), dictates the density of our bones. There are other genetic influences: for example, white women are two to three times more likely to fracture their hip than non-white women. This is because negroid adults are destined to have a greater bone mass than Caucasian adults. Differences in bone density can be picked up before puberty and correlate well with our inherited sensitivity to vitamin D.

Physical activity will reduce the likelihood of developing osteoporosis in later life, whereas immobility accelerates its progress. One dramatic case is that of astronauts who have spent a prolonged period of weightlessness; they are particularly vulnerable to osteoporosis. Bone density improves some time after they have come back down to earth and reintroduced their bones and muscles to the stress of carrying the body's weight at normal gravity.

People with severe osteoporosis have thinning of the vertebrae of the spinal column. When the vertebrae begin to collapse down an individual loses height: a common sight is the little old lady with 'dowager's hump' – bent double with osteoporosis of the backbone.

As if they don't do enough harm to the body elsewhere, smoking and alcohol also increase the rate of bone loss. However, sex hormones are most important in determining our bone mass. Starting around the fourth or fifth decade of life, both men and women lose bone at a rate of 0.3 to 0.5 per cent per year. In women, a sudden fall in oestrogen levels after the menopause increases this bone loss, by up to 10 times, for the next 10 years. After that period they continue to lose bone at the same steady rate as men. Men's rate of bone loss will increase like women's only if the level of their sex hormones falls drastically (for example, after castration).

Measures taken to slow down the inevitable loss of bone in later life include an increase in physical exercise, additional calcium and vitamin D in the diet and hormone replacement therapy (HRT) for women. As bone loss in women is most rapid in the first 10 years after the menopause, HRT is most effective at preventing bone loss if given at this time. There has also been success in slowing the progression of severe osteoporosis with a group of drugs that inhibit the action of osteoclasts (that is, prevent the bone being destroyed) – the bisphosphonates. With all these efforts to prevent osteoporosis and consequently reduce the chance of a fractured bone, another effective strategy should be borne in mind: creating a safe environment for a vulnerable person and providing suitable aids to mobility. Simply not falling over may be a lifesaver.

newly formed bone equals the amount of old bone destroyed. In childhood, this balance favours new bone formation, while in later life bone destruction is predominant. When excessive bone destruction occurs, osteoporosis develops.

In the late twentieth century, throughout the world, men and women are living longer. With this longevity, however, comes an increasing tendency towards physical disability. One of the main causes of this disability is osteoporosis. It is a disease in which the bone-eating action of osteoclasts outstrips that of the bone-forming osteoblasts. The overall density of bone is therefore reduced. The great danger of osteoporosis is that the fragile bones are more vulnerable to fracture. Osteoporotic fractures occur in 50 per cent of women and 30 per cent of men over the age of 50. Each year in the USA more than 250,000 people fracture a hip: 90 per cent of these people are over 50 years old. The risk of fracturing a hip doubles for each decade after 50 years. Women after the menopause are the most common victims. As the population continues to age, the annual number of hip fractures is projected to double by the year 2040. This creates an ever burgeoning stress on the carers of the increasing number of elderly and disabled people.

On average, around 10 per cent of a body's bones are replaced every year. Over time this process slows down but, even if we live to be 95 years of age, our bones will never be much older than ten.

## GROWING UP

We grow as cartilage is turned into bone, a long, drawn-out process known as ossification. It begins in the womb, when the foetus is around 10 weeks old. The developing foetus is entirely dependent on its mother's system for the calcium it needs; after birth, if it is breastfed, the baby continues to take huge quantities of calcium from the mother's milk. This calcium helps mineralise the baby's skeleton, but at the expense of the mother's bones. After six months of breastfeeding, women reduce their bone density by around five per cent – more quickly than after the menopause. This loss of bone density quickly recovers once breastfeeding stops and appears to have no long-term consequences. Interestingly, several studies have now shown that supplementing the mother's diet with calcium does not prevent the loss of bone density during breastfeeding. As yet there is no answer to this problem, but research is continuing.

The newborn baby grows very quickly: during the first three months of life it will grow by an average of three centimetres per month. Over the first 12 months the child will grow an average of 25 centimetres. The rate of growth then decelerates to about six centimetres per year, until puberty; ossification continues until we reach our adult height.

In long bones, such as the radius in the forearm, the shaft of the bone is ossified but the ends of the bone remain lined by a specialist layer of cartilage that becomes the surface of the joint. Underneath the joint surface is another layer of cartilage that persists as the growth plate or growth cartilage. The growth cartilage allows an increase in bone length by pumping out new bone until the full adult dimensions are reached.

When bones have reached their final length, the growth cartilage becomes thin and ossifies. The mature bone is formed – and in a natural lifespan it will be completely replaced at least seven times.

## The role of vitamin D

The proper rate of growth is controlled by the correct cocktail of growth hormone, secreted from the pituitary gland, and vitamin D. The main job of active vitamin D is to increase the absorption of calcium from the intestines, making more calcium available for the mineralisation of bone. Throughout our lives, the requirement for vitamin D varies. It plays a vital role in the growth of children's bones, the maintenance of adult bones and prevention of osteoporosis in elderly bones. A lack of vitamin D in children leads to rickets, which in adults – who have fully grown bones – is known as osteomalacia (see page 40). Recently, one large study found that more than half of all hospitalised patients were deficient in vitamin D.

Although vitamins are essential for normal health, it does not follow that excessive amounts provide good health. Some people feel that swallowing a large, bitter multivitamin tablet absolves them from the effort of healthy living. Yet it does seem that many of us are not taking enough vitamin D. It's found in great abundance in fish livers, especially cod (hence the punishing prescription of codliver oil to children). Nowadays, large amounts of vitamin D are found in fortified foods such as cereals, milk and infant feeds, as well as eggs and butter.

Riding a bicycle involves the balance centres in the inner ear linking with visual information from the eyes and a complex system of muscles, tendons and ligaments in the arms and legs. Faliure of any one of these systems can have catastrophic results.

## Rickets

In 1650, when a diet deficient in vitamin D was commonplace, Francis Glisson (1597–1677) first accurately described the physical appearance of rickets. Glisson, Regius Professor of Physic at Cambridge, was unaware of the underlying causes of the condition, but noted the visible evidence: delayed growth, weakening and bowing of weight-bearing bones, and thinning tooth enamel. Later scientists discovered the relevance of a low calcium level in the blood, with consequent weak muscles that have a tendency to go into spasm (tetany).

The adult form of rickets is osteomalacia. As the growth plates have stopped working in adults, the clinical appearances of osteomalacia are much less dramatic than those of rickets.

At the beginning of the twentieth century, rickets had become a widespread problem in northern industrialised nations. For example, at this time in Boston, USA, it was estimated that 80 per cent of children under two years had rickets. The solution to this massive problem was to add vitamin D to the diet.

However, vitamin D is still absent from the diet of millions of undernourished children in poor countries, but rickets is not as great a problem as one would expect. This surprising observation was first made late in the last century, when the prevalence of rickets in children from poor homes in grey, northern cities was noted. Exposure to sunlight was recommended as a treatment for rickets – and this proved a most intuitive prescription.

It was not until the first half of this century that the role of sunlight in the prevention of rickets and osteomalacia was discovered. Amazingly, within our skin there is a molecule which, when exposed to the ultraviolet light of sun, is converted into vitamin D. If we are not exposed to sunlight, then this important source of vitamin D is unavailable to us. Combined with a diet deficient in vitamin D, people from countries with poor sunlight are vulnerable to rickets and osteomalacia. At least this is not another problem to add to the plight of those struggling in countries with limited food resources but abundant sunlight.

Vitamin D needs to be activated by the kidneys before it can carry out its job. If a patient has kidney failure, then vitamin D remains inactive and the patient suffers all the complications of a person with a diet low in vitamin D, such as rickets or osteomalacia. Nowadays, patients with kidney failure can be given tablets containing activated vitamin D.

### Size isn't everything!

Assuming that a child does receive the correct balance of hormones, calcium and vitamin D, the height it has reached on its second birthday is a good indication of its future growth; by this age it will be roughly half its final adult height. However, the influence of maternal height is strongest in the first three years of life, whereas paternal height has a greater influence on the growth pattern in later childhood. Large discrepancies in the parents' height can therefore give unusual growth patterns. Like plants, most children grow more quickly in the spring and summer, with slower growth in the winter and autumn. The pubertal growth spurt begins two years earlier in girls than boys but, by the age of 14 years, the average male will have caught up with his female counterparts and will now be taller. In the UK, the average adult male is 1.78 metres and the average adult female is 1.64 metres.

## Growth disorders

Growth hormone, secreted from the pituitary gland at the base of the brain, is essential for normal growth. Rarely, too much or too little growth hormone can be produced by an abnormal pituitary gland. Too much growth hormone in childhood leads to gigantism, where the child grows far above the expected height for its age, well beyond 'normal' tallness. The most extreme case recorded so far is that of a young man in Illinois, USA, who died in 1940 at the age of 22. Just before he died he measured just over 2.72 metres.

Too much growth hormone in adults, after the growth plates have ossified, leads to a condition called acromegaly. Here, instead of bones growing longer, they become thick, especially the skull, cheekbones and jaw. Hands and feet become wider, so that rings have to be cut from fingers and shoe size goes up. The skin also becomes thicker, and so do the vocal cords, making the voice deeper. The treatment of this condition is either surgical removal of the abnormal pituitary gland, or medication to reduce the level of growth hormone.

Children with too little growth hormone grow at a slow rate and are far below the expected height for their age. Special tests are required to measure growth hormone levels; if levels are low,

then an magnetic resonance imaging scan of the pituitary gland is necessary to look for any abnormality. Medical or surgical treatment may be required to treat any pituitary abnormality, and genetically engineered growth hormone can be given to replace any deficiency. Before this type of hormone was available, it used to be obtained from extracts of animals' pituitary glands. Unfortunately, many young children developed a condition known as Creutzfeld Jacob disease (CJD), due to unidentified infection within the animal growth hormone. A new variant of CJD has now also been blamed on the same infection, caught from eating affected beef.

Sometimes the growth plates at the end of long bones fail to work, leading to short arms and legs. Short limbs associated with a normal-sized skull and trunk are typical of a condition called achondroplasia (more commonly, dwarfism). This is usually an inherited condition, but many cases occur out of the blue. Children with achondroplasia are of normal intelligence. The main complications of the condition are related to the bone abnormalities, which predispose to osteoarthritis later in life. As yet, there is no known way of stimulating the growth plates of the long bones into action.

However, there is a wide range of what might be called 'normal' height. In an age where the belief most widely held is that 'tall is beautiful', some people might be anxious if they seem to fall short! This is perhaps more pertinent for men, rather than women, as there is greater social pressure put on a man of below-average height. Technically people are defined as 'short' simply if they fall below the third centile for their age. Put another way, this means that at least 97 per cent of the population are taller than them. There are many reasons for short stature, the most common being that both parents are short. Obviously little can be done about this, although there is some evidence that genetically engineered human growth hormone can accelerate growth in the children of short parents. Even so, current studies of children prescribed this growth hormone have shown little difference to the final adult height. However, despite this evidence, patients and parents alike can still exert powerful pressure on the prescription of growth hormone, although some physicians feel that the ethics of treating healthy young children with this expensive drug are dubious. It is only when growth patterns are abnormal that medical intervention is thought to be justified (see Growth disorders above).

## Common fractures

One of the most common kinds of broken bone is the Colles fracture. This is named after Abraham Colles, a Dublin surgeon, who wrote a remarkably accurate report on injuries of the wrist in 1814–80 years before the discovery of X-rays.

It affects one of the bones in the forearm, the radius, close to its junction with the wrist. It classically occurs after a fall on to the outstretched hand. The hand is pushed backwards, forcing the radius to snap, so that the lower end of the bone rides over the upper end. The hand and forearm are temporarily deformed, giving a typical 'dinner fork' appearance. In order to correct this deformity, the radius bone must be manipulated out to its full length again. This can be done under local or general anaesthetic, before putting the hand and forearm in plaster. The arm is then X-rayed a week later, to make sure the position hasn't slipped back again. If it has, the broken arm needs to be reset.

A common kind of broken bone in children is called a 'green stick' fracture. When the green stick of a plant breaks, it does not snap cleanly. One side may break, while the other side buckles. When the pressure leading up to the fracture is released, a green stick doesn't quite return to its original position but springs back into a bent position. Fractures in the long bones of young children behave in the same way. Children's bones have a much greater capacity to bend than adult bones. This is a mixed blessing: although the bones remain in continuity, their springy nature does not allow the fracture to be manipulated back into a perfect position. Fortunately, children's bones remodel very well to restore normal anatomy.

## JOINT PARTNERSHIP

Two bones come together at a joint. Depending on the shape, size and function of the adjacent bones, the design of joints varies enormously. The most basic design would not be considered by many people to be a joint at all. For example, the skull is made of several different bones, widely spaced during infancy (to allow the brain to grow unhindered by a fixed size of skull), but fused at a so-called 'fibrous joint' during adult life. Another example is the fibrous joint between the teeth and jaw bone. Normally, these joints allow no movement.

A slightly – but only slightly – more mobile joint is that between the vertebral bones of the spinal column. The vertebrae are separated by a disc of cartilage, which allows a degree of movement within the spine. Occasionally, after heavy lifting in an awkward position, this disc of cartilage can be forced out between two vertebrae: a 'slipped disc'.

The most common joint design allows a wide range of movement. The ends of the bones are covered by cartilage and separated by a tiny gap – the joint cavity. A small amount of lubricating fluid is produced by a membrane that lines the joint. This fluid also nourishes the cartilage, which has no blood supply of its own. Cartilage wears down with age and is therefore a feature of 'wear-and-tear arthritis' – osteoarthritis. Unlike bone, cartilage does not show up on an X-ray. If an X-ray shows a narrow space between two bones, it suggests that the cartilage has been worn down and osteoarthritis is likely.

Two bones at a joint are held together by ligaments. These tough fibrous bands of tissue prevent dislocation of the joint. A torn ligament can be more serious than breaking a bone. It may need repairing or even replacing but, if the tear is not too serious, a period of immobilisation is enough to allow the ligaments to repair themselves.

In order for bones to move at a joint, they must be attached to muscles. The ends of muscles become fibrous just before they attach to bone and at this point are known as tendons. One of the largest tendons in the body joins the calf muscle to the heel – the Achilles' heel.

Throughout our lives, our joints are put under huge stresses. To take one simple, commonplace example: every time a child hops, her ankle has to withstand a force 10 times her body weight. Our leg joints are subjected to greater or lesser stress every day of our lives. Such stress will go through a joint around 50 million times, without the joint ever needing to be replaced. The stress inflicts repeated micro-fractures; these must be mended at the same time as the cells are ceaselessly renewing ageing bone.

Sometimes the human body is required to do more than a gradual repair and renewal job. Sometimes it has to deal with a major catastrophe, when we are involved in a violent, painful accident.

## BREAKDOWN OF AN ACCIDENT

We can take as a typical example a young girl falling off her bicycle, something that happens somewhere every day. She is still learning the skills needed to perform her daily activities – skills that we take for granted by the time we have grown. Cycling requires the brain to instruct individual muscles to tense and relax, while filtering signals from the balance centres in the inner ear. We make subtle adjustments to keep our body balanced. It is a system of breathtaking complexity involving practically every bone and muscle in our body. Sometimes this integrated pathway fails us and we fall – even as adults we can make mistakes, like misjudging distances, so it's all the more challenging for a child.

Riding a racing bike over rough terrain can increase the likelihood of an accident.

**Right:** When we fall our natural instinct is to put out an outstretched hand. The bones at the wrist are therefore most vulnerable to fracture. In a typical fall, the force transmitted into the radius is 25 times our body weight. This is too much for the resilient radius bone to bear, so it breaks at its weakest point.

**Overleaf:** A fracture in the shaft of the radius is painful but less damaging than a fracture through the wrist or elbow joints.

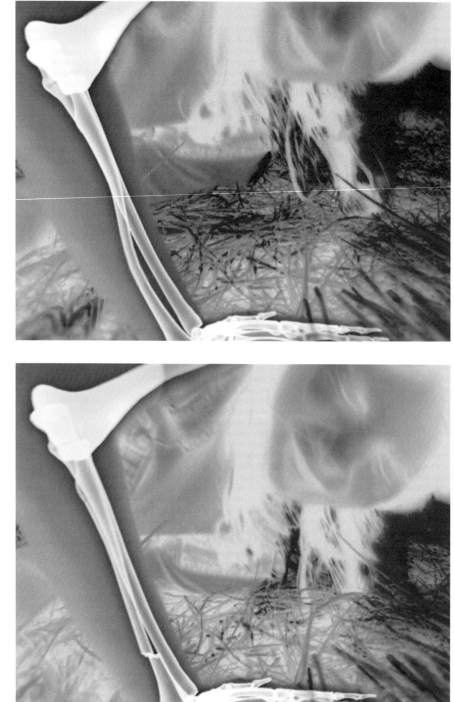

When we fall another integrated system is triggered. This time it is our subconscious brain that instructs our movement. When we put out an arm to protect our body, the sudden force transmitted through the radius bone in our arm is 25 times our body weight. Despite the bone's incredible strength, under this much pressure the narrow part is designed to act like a fuse. It snaps in order to prevent damage to the rest of the body. So the young girl falls heavily and awkwardly, breaking her arm and getting covered in cuts and bruises.

## Bone damage and repair

Inside the bone, the scene is one of carnage. The intricate arches have been snapped to pieces. Hundreds of tiny blood vessels have been severed. Battered pain nerves send frantic signals up her spinal cord to her brain. Her brain responds by releasing its own natural morphine – endorphins, which mute the pain signals to almost nothing. The brain evolved endorphins to keep pain at bay for a few vital minutes in case we need to escape danger. Initially, the wounds are all anaesthetised: within 30 minutes, the endorphins will have worn off. (There are many reports of soldiers fighting on after losing a limb, becoming aware of their wounds only when the fighting has calmed down.)

The pain is unpleasant, but it is also essential. It is our body's way of alerting us to internal damage. As the pain grows, our instinct is to rub near the site of injury. This rubbing is another natural form of pain relief, the action releasing another wave of pain-blocking chemicals. But the effect is much milder than the endorphins, much too mild to counter the frantic signals coming from a broken bone.

I can speak from personal experience. While still working on this chapter, I myself fell off my bicycle. I'd been forced to brake sharply by a wayward car and flew over the handlebars. By the time I got back on my feet, I realised I had cuts and bruises but otherwise felt fine. About an hour later I became aware of pain in my right chest. It was tender and hurt when I took a deep breath. An X-ray showed a fracture of my right tenth rib. There was fortunately no punctured lung, which can sometimes occur when a rib is fractured. The relevance here – apart from the coincidence of having my own accident while writing about someone else's – is the delay in pain from the fractured rib. As I pedalled away from the scene, the endorphins were roaring round my body and I had no idea of the damage to my rib. When the pain did kick in, no amount of rubbing made me feel better!

While the girl may not feel pain for a while, her body is already showing

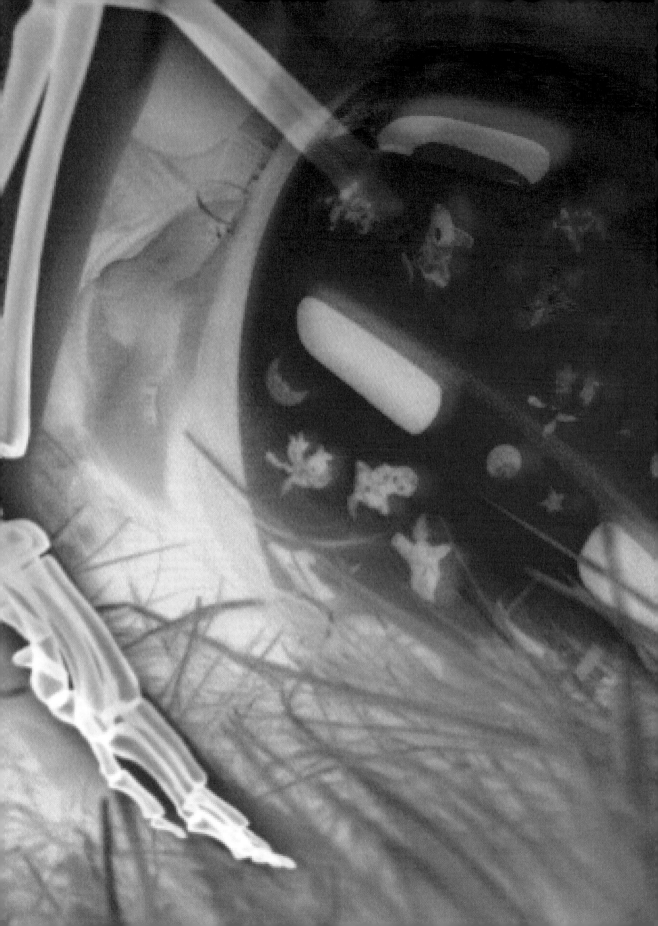

## Brittle bones

While fractures are usually caused by external forces, there is a rare bone disease that predisposes patients to fractures. This is known as brittle-bone disease or osteogenesis imperfecta, and it can be fatal. It affects about one in 20,000 births and occurs in different forms. The most severe form leads to multiple fractures at the time of birth, due to poor ossification of the bones. Collagen production is also reduced, which is apparent not only in the brittle bones but in other tissues too. Most notably the whites of the eyes – the sclera – appear blue. The thin white collagen of the sclera becomes transparent, allowing the blue pigment of the underlying tissue to become visible. Less severe forms of the disease still bring about multiple fractures in childhood; repeated fractures lead to stunting of growth and deformed limbs. This often leads to the mistaken diagnosis of 'battered baby syndrome', especially if associated with clinical signs of neglect. The distinction can be difficult, but has important legal consequences.

There is no convincingly effective treatment for brittle-bone disease. Bone density may be improved with bisphosphonates – drugs that prevent the bone being destroyed – but there is no evidence that the number of fractures is reduced.

Bones affected with cancer also become fragile and vulnerable to fracture. As the bone is severely weakened, these fractures often occur with only minimal stress and are frequently painless. Consequently the patient does not know they have a fracture until a deformity is obvious. Radiotherapy to the fracture site can help the fracture unite and reduces pain if present.

remarkable signs of its ability to heal itself. Her cuts and grazes are bleeding from tiny severed vessels, and this must be stopped quickly. Specialist blood cells called platelets change their shape as they come into contact with damaged blood vessels; they become sticky and lock together to plug the wound. At the same time, proteins circulating in the blood solidify into strands of a substance called fibrin. A net of fibrin is thrown across the wound, trapping the blood cells underneath. The fibrin net contracts, squeezing liquid out of the clot like water out of a sponge. And as the clot shrinks, it performs one more miracle. It pulls the skin together, closing the wound. At first this is a delicate process, that can be blown apart by clumsy but essential efforts to clean the wound. As soon as the wound is clean and dry it will swiftly start to heal again.

Underneath the skin the bone has called in its own emergency services. The tissues around a fractured arm, battered by a fall, send out chemical signals which make the arm double in size. This swelling is caused by blood vessels in the arm dilating. Fluid passes from the blood into the surrounding tissue, flooding the fracture site with oxygen and other nutrients. The chemicals that trigger swelling also cause the pain nerves in the arm to become more sensitive. The slightest pressure has them firing off pain signals to the brain (the endorphins having worn off by now). The nerves are so sensitised they can feel the blood pulsing through swollen vessels – a throbbing ache. But again this pain has a purpose. It protects the arm by making sure nothing is done to aggravate the injury.

This girl's accident is fairly minor; it could have been much more serious. In more severe cases, the wounds can tear through skin and muscle, exposing the underlying broken bone. This is known as an open or compound fracture. In this situation, there is a serious risk of bacteria entering the site of the fracture and causing an infection in the bone. It is always assumed infection is lurking, so after the wound is thoroughly cleaned with antiseptic solutions, a course of antibiotics is given and possibly a tetanus booster.

One and a half hours after the girl's accident, things have moved on. Just as on the skin, blood is clotting inside the broken bone. Depending on the number of blood vessels severed, it can take several hours before the flow of blood within a fracture stops. The result is a huge blood clot, called a haematoma, enveloping the entire fracture.

By now the girl will have been taken to the casualty department of her local hospital, where her arm will be X-rayed. Then the broken arm will be put in plaster to prevent the fractured ends moving about and causing more damage. However, modern medicine can do very little to accelerate the actual healing of a broken bone. It is the bone's own self-healing processes that must do all the work. (In the case of a broken rib, like mine described on page 45, nothing can be done, not even plaster or bandages.) Once the haematoma around the fracture is fully formed, the body performs a miracle: it transforms a massive clot from blood to bone.

At the point of fracture the scene is one of carnage. Torn arteries and veins bleed into the damaged area causing a haematoma (bruise). Frantic nerve signals are sent back to the brain, and the intricate arches of bone have been shattered.

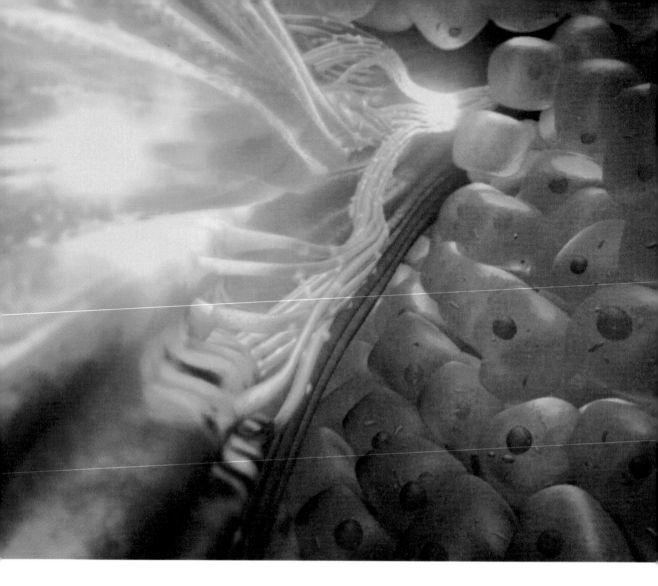

Pain signals travel relatively slowly along the nerves and can be blocked by other nerves that carry the faster-moving sensation of rubbing.

To mend a broken arm, thousands of osteoblasts – the bone-building cells – are urgently required. The spaces deep inside the bones are inhabited by stem cells, which normally divide to create a new osteoblast once every two days. When a bone breaks, the stem cells must divide once every three minutes, to create an army of bone-building osteoblasts.

Within eight hours of an accident, the osteoblasts are inside a haematoma. The minerals they release enclose the clotted blood cells in tough bone. Outside the clot more osteoblasts are working: the transformation from blood into bone is under way. The body is doing what it was designed to do supremely well – repairing itself, using its own resources.

Repair work is also continuing elsewhere in the girl's body. When her knee hit the ground, the skin was grazed in some places but not in others. Underneath areas of intact skin some blood vessels were severed, leaking blood which cannot escape. The trapped blood spreads out underneath the skin, until the torn blood vessel is repaired. The iron-based pigment, haemoglobin, which gives blood its red colour, is gradually broken down chemically. This pigment changes colour as it breaks down, displaying the characteristic rainbow hues of a bruise.

One week after the accident, the clotted blood within the grazes has dried up to form a scab. New skin grows at a remarkable rate around the scab. A millimetre beneath the skin, stem cells are rapidly dividing to create new skin cells. The new cells are pushed up to replace the damaged skin above. This process loosens the scab, which can then be pulled off. Children heal far more quickly than adults, laying down new skin in such a hurry that there's no time to replicate the pattern of the original skin. Unfortunately, this means that children also scar more easily than adults.

Meanwhile, the two ends of the broken bones steadily bond together. The plaster cast protects the injury, but it is important not to restrict the movement of

**Overleaf:** Once the arm is fully healed it is impossible to see where the fracture occured.

## Tetanus

Tetanus is a nasty infection caused by a bacteria known as *Clostridium tetani*. This organism produces a powerful toxin which targets the nervous system, causing muscle spasm. If the muscles that control breathing go into spasm, then the patient can die. The tetanus bacteria is usually found in soil and animal faeces, but has occasionally been found in places as supposedly clean as operating theatres!

Tetanus is a disease that mainly affects the poor, underprivileged people of the world, especially the young. Worldwide, about 800,000 newborn babies are killed by tetanus each year. In 1992 the World Health Organisation found that in most developing countries tetanus accounted for about half of all baby deaths. It is a much rarer disease in the western world, mainly because of an effective immunisation programme. In the USA, there are only about 70 cases a year, usually affecting people over 50 years of age, whose immunisation has worn off.

In the western world, infants are given a triple vaccine that protects against tetanus, diphtheria and whooping cough. This triple vaccine is given on three occasions within the first year of life. The tetanus part of the vaccine is made from an inactive version of the bacteria's own toxin. The immunised child then makes antibodies against the tetanus toxin, ready to be put into action if infected. This is known as active immunisation, as the child's own immune system is activated to produce antibodies in response to a toxin or infection. After such an immunisation programme, protection against tetanus lasts for many years, but can be enhanced by a booster dose around five years of age. Further booster doses should be given every 10 years.

Often, the opportunity to give a tetanus booster arises only at the time of a minor injury. A severe or infected wound should be cleaned as usual, but if more than five years have passed since the last tetanus booster – so the patient's immune response is starting to wear off – then an injection of ready-formed antibodies should be given to neutralise the effects of the tetanus toxin. This is passive immunisation, where the patient's immune system does not itself actively produce any new antibodies.

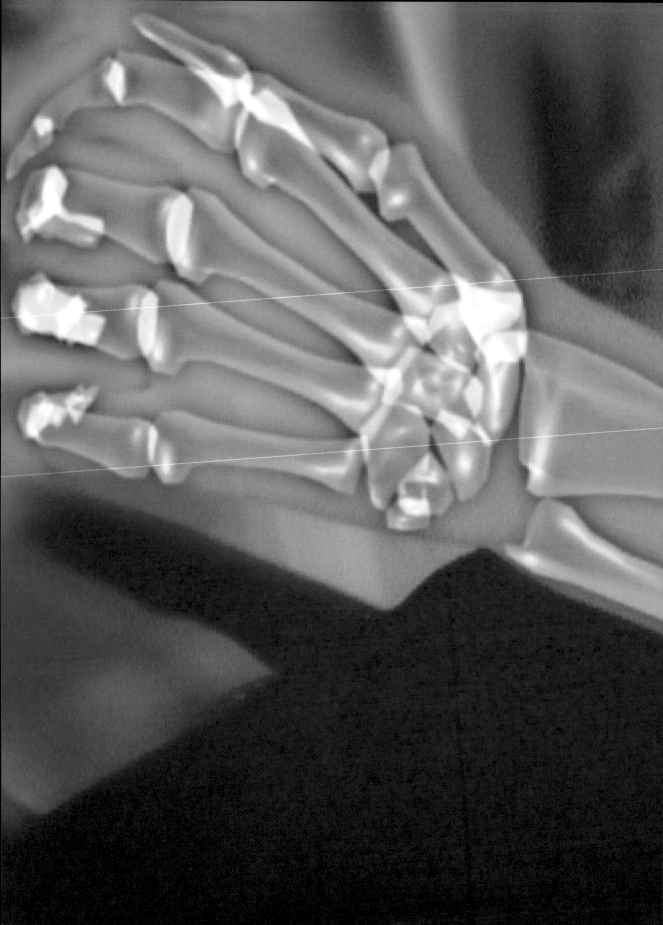

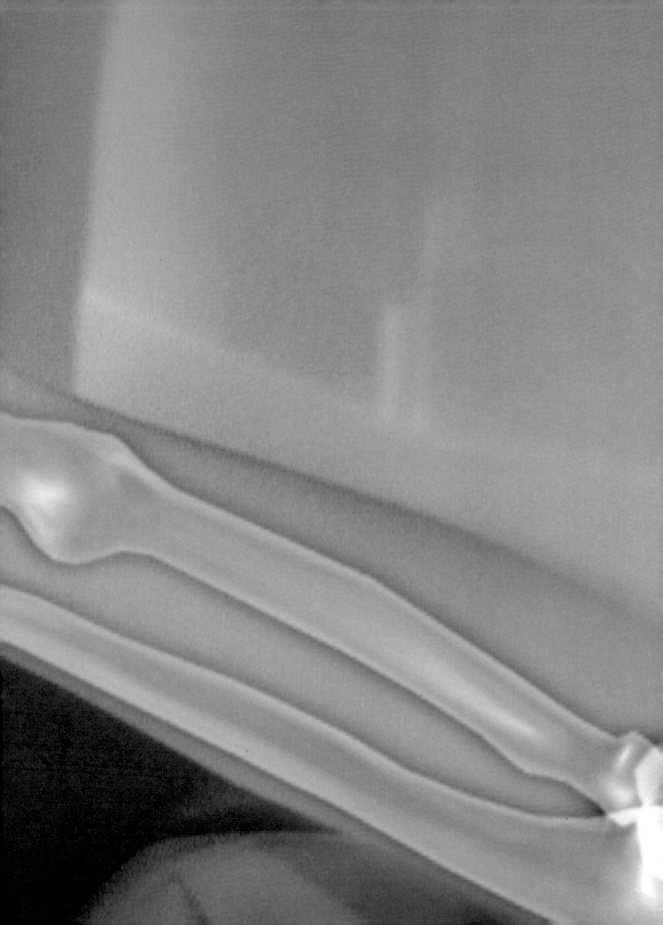

**Right:** Broken bones are put in plaster to stop the broken ends from moving about too much. But a little movement is a good thing as it promotes healing. The bone-repairing osteoblasts are stimulated into action by tiny electric currents generated by the movement. Happily for growing children, their bones heal twice as quickly as adults' bones.

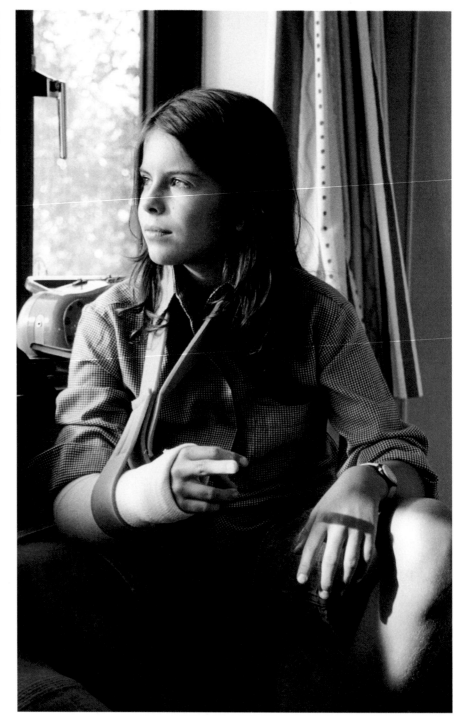

the bones completely. Movement of the fingers makes the two ends of the break shift gently in relation to each other. This movement generates tiny electric currents, which seem to invigorate the work of the bone-building osteoblasts. Movement is actually helping to fix the broken bone.

Until the osteoblasts have finished their work, the arm can still be damaged by external impact. Again, pain is the body's best way of making us protect vulnerable areas.

Four weeks after her accident, the plaster cast is ready to come off. Deep inside, the haematoma is almost pure bone. The bone-builders are finishing their work, and in the process something quite remarkable is happening. The bones of growing children heal twice as fast as those of adults. As we get older, our bone's regenerative powers decline, but with time they can always perform their own miraculous repair.

The haematoma has been transformed into a solid ball of bone, twice as thick and strong as the original bone which was there. But having a stronger section of bone could be a handicap. If the girl falls over again, the radius might not break in the way nature intended, and more serious damage could be done. So the osteoclasts – the bone-destroyers – stir into life. They begin dissolving away the thick, over-repaired bone with jets of acid. They will continue to resculpt the bone until it is exactly the shape it would have been, had it grown normally and never been broken. Although it has taken only a month to fix, the resculpting process will take another whole year. The bone-destroyers will have carved tunnels through the inside of the bone, perfectly restoring its intricate arches. It will be filled with blood vessels and teeming with life again. It will look like any other bone in the body, as if the accident had never happened.

## Looking beneath the surface

On 8 November 1895 Wilhelm Rontgen (1845–1923), a physics professor at Wurtzburg, discovered *eine neue art von stralen* (a new kind of ray). He took the first X-ray (so called because of the unknown source of the ray), of his wife's hand. Soon the technique was used for examining broken bones, but initially the exposure time was very long, up to 30 minutes, and caused the skin to burn. Later, this ability of X-rays to burn skin was put to good therapeutic use in the treatment of skin cancer.

Today our ability to look inside the body has advanced tremendously. Not only has the exposure time for X-rays fallen to a split second, but the total dose of radiation from a single chest X-ray is no more than that experienced on one transatlantic aeroplane flight. Modern techniques such as computerised axial tomography (CAT) scanning and more recently magnetic resonance imaging (MRI) allow far more detailed examination of not just the bones, but the soft tissues within the body.

There are several ways in which the density of bone can be measured. The most widely used is a dual X-ray absorptiometry (DEXA) scan. This measures the amount of calcium within a given area of bone and is expressed as bone mineral density (BMD). The BMD level can be compared with the average for the appropriate age, as the density of bone falls as we get older.

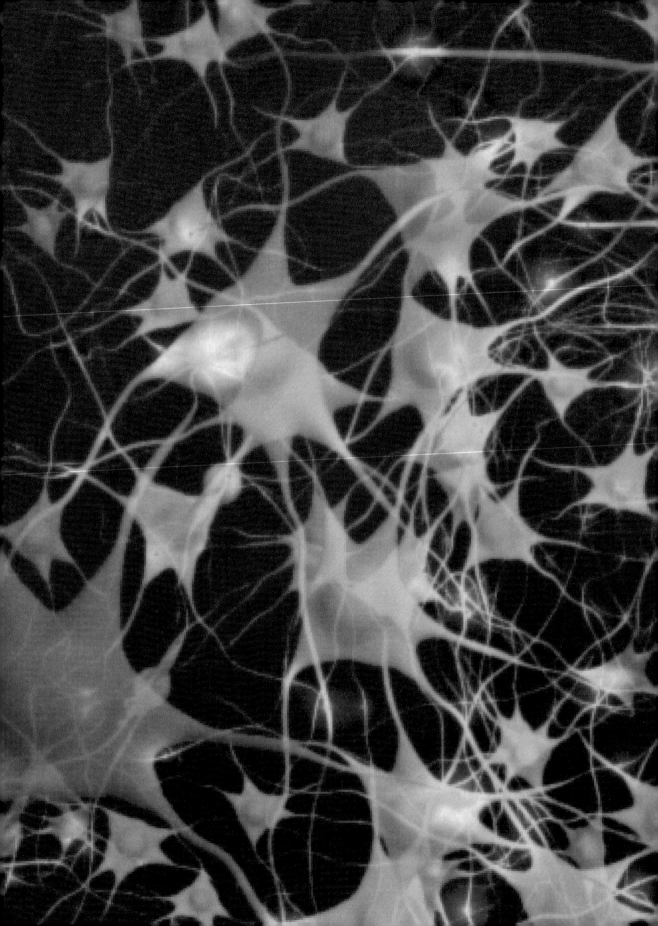

# 3
# BASIC
# INSTINCT

Controlling every single aspect of our bodies, the human brain is the most complex piece of organic machinery in the universe. Behind our rational thinking brain is a more primitive brain controlling the essential drives that further our survival: our basic instincts.

**Previous page**: When we
are born the 100 billion
nerves of the brain lack all
the trillions of connections an
adult later develops. This
explains the endearing but
clumsy physical nature of
newborn children.

**Right**: Deep below the folded
grey matter of the cerebral
cortex sits the hypothalmus
which controls almost every
hormone in the body.

While we can consciously order our thoughts and actions, much of how we function is unconscious: our biological systems have evolved so that vital processes such as breathing, heartbeat and hormone production are carried on automatically. You don't have to consciously decide how many times a minute your heart should beat, just as you don't intellectually consider your body's physical 'fight or flight' response to danger.

Over millions of years of evolution, the ancient, primitive brain – the 'animal' brain of instinct – has been augmented by the 'higher' brain, which gives us our intelligence, our rational thought and our sophistication. It is the whole brain that controls all our thoughts and actions, whether conscious or unconscious. Philosophers and scientists have struggled for generations to understand how such a physical structure as the brain can contain such an ethereal entity as our minds. Most now believe that our mental processes are the result of physical activity within the brain.

## MIND OVER MATTER: THE CEREBRAL CORTEX

The brain sits at the top of the nervous system, both literally and figuratively, a delicate structure with the texture of a hardboiled egg. It looks like a giant walnut but is unimaginably more convoluted – it is made up of about 100 billion nerves linked to each other through at least 10 trillion connections. Removed from the skull, the brain weighs about 1,500 grams but, floating in the fluid within the skull, it weighs only 50 grams. This liquid buffer prevents the important structures on the underside of the brain from being squashed by its own weight.

With some justification, we might regard ourselves as the most intelligent of all animals. However, we do not have the biggest brains – an elephant's is five times the size of ours. Nor do we have the biggest brains relative to our body size, a distinction probably earned by the shrew (an animal not noted for its intellectual flair). The most striking difference between our brains and that of other animals, including the chimpanzee (with which we share 99 per cent of our genes), is the size of a certain part of our brain: the cerebral cortex.

The cerebral cortex is the outer layer of the brain ('cortex' comes from the Latin for bark). If we examine the outside of the brain we can clearly see it is divided into two halves, or hemispheres, and is then folded into four major lobes (frontal, parietal, temporal and occipital). Further multiple folds within each lobe greatly increase the surface area of cerebral cortex relative to its volume. It

## Matter over mind

The brain is a solid structure without any visible moving parts, yet it controls our thoughts and actions. Before the development of modern methods of analysis, scientists could look at the brain only when it was dead, and consequently could only guess at the way it worked.

One favourite theory reckoned that the underlying workings of the brain could be established by examining bumps on the skull. Led by the Austrian, Franz Gall (1758–1828), this piece of quackery was later known as phrenology (from the Greek for 'study of the mind'). Phrenology was soon discredited when clinicians performed post-mortem studies on the brains of people who had

neurological disease. They found no correlation between the bumps on the skull and the proposed personality traits outlined on phrenology busts.

Modern technology has now advanced so that we no longer have to wait until the patient is dead before analysing the brain. Special X-rays – CAT scans – can take images of the brain as if it were sliced at several levels. But, even better images can be obtained by magnetic resonance imaging (MRI scans), which doesn't use X-rays but rather radio-frequency energy in a magnetic field. The function of the living brain can now also be observed using a special adaptation of the MRI scan.

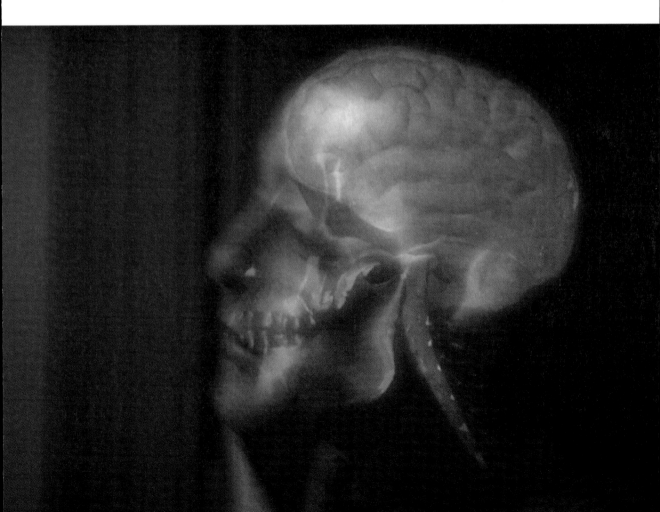

justifies the common expression 'grey matter', being packed full of billions of cell bodies that give it a grey appearance. It is two to four millimetres thick and makes us both clever and dangerous. It is the seat of our intelligence and sophistication, but has also endowed us with a tendency to kill each other, unique throughout the animal world.

All our five senses – sight, sound, taste, smell and touch – are relayed by nerves to different parts of the cerebral cortex. This information is instantly assimilated and integrated before being relayed back down to other parts of the nervous system for action. It has taken millions of years for our brains to evolve such a sophisticated cerebral cortex, but only in the last 40,000 years have we really begun to exploit its potential.

The capacity for language was a major evolutionary breakthrough. The necessary development of vocal cords was accompanied by a large and specialised brain capable of understanding speech and then instructing the voice box to articulate the spoken word. Initially, language was probably only a replacement for the prolonged grooming sessions performed by males: an act of flea and lice picking that is still performed by male chimpanzees. Such behaviour is much prolonged when males are in the company of females on heat, and could therefore reflect a form of sexual bargaining. Only since the development of anatomically modern humans, about 100,000 years ago, has language been used for communicating information. At first it enabled valuable skills learnt by the elders in the community to be passed on from generation to generation; later, written language could be used in the same way, this eventually lead to the exponential speed of technological development we are familiar with in the modern world. Despite this rapid pace of development, we are all almost certainly, still underusing our brains.

In almost all people, one half of the brain dominates over the other, by virtue of possessing the major speech areas. It used to be thought that the dominant hemisphere was always on the opposite side to that of the preferred writing hand. In over 95 per cent of right-handed people the left hemisphere is indeed dominant, but it is also dominant in over 70 per cent of left-handed people. However, the right cerebral hemisphere also has its strengths, in particular the visual skills that recognise three-dimensional images. This may perhaps explain why those 30 per cent of left-handers with a dominant right hemisphere appear to be more likely to be skilled artists.

## Mapping the brain

In 1908 a neuro-anatomist, Korbinian Brodmann, noted slight differences in the arrangement of nerves in different regions of the cerebral cortex. He mapped out 47 separate regions that were later shown to have different functions. For example, area 44 in the frontal lobe of the left hemisphere corresponds to verbal fluency. It was actually identified by a French neurologist, Paul Broca, almost half a century before Brodmann drew his map of the brain. In 1861 Broca performed a post-mortem on a man who in life had normal understanding of words, but a frustrating inability to articulate his thoughts. The patient became known as 'Tan', named after the only utterance he could make in response to questions. The post-mortem revealed damage to a part of the brain now known as Broca's area. Tan also had a paralysed right arm, indicating that part of the left hemisphere, controlling movement (next to Broca's area), was also damaged. The most likely cause for such brain damage is a stroke (cerebro-vascular accident), caused by either a haemorrhage or a clot in the blood vessel supplying that area of brain.

Another remarkable case which contributed to our understanding of brain function was that of Phineas Gage. In 1848 this foreman of a small working party on the US railway was caught in an explosion. The iron bar he used to pack down explosives flew through his cheek and out through the top of his head. Despite horrific damage to the frontal lobes of his brain, he survived. However, his character changed beyond all recognition. Before the accident he was a kind, considerate man who was highly conscientious at work, but afterwards he lost all his inhibitions and became loud, unreliable and easily distracted.

This accident, and similar observations in other patients, suggested that destruction of the frontal lobes left a patient alert and orientated, with a normal memory and intellect, but with a much reduced attention span and limited ability to plan ahead. Despite these disadvantages, an operation to disrupt the nerve fibres between the frontal lobes and the rest of the brain, a frontal leucotomy (often known as a frontal lobotomy), was performed in the 1940s and 1950s for the treatment of anxiety and depression. Not until more than 10,000 frontal leucotomies had been performed did neurosurgeons and psychiatrists realise that the consequences of this operation were usually worse than the original symptoms.

# LIFE IN THE FAST LANE: THE PRIMITIVE BRAIN

Information from the cerebral cortex passes to other, more primitive parts of the brain via nerves with a characteristic white coating – the white matter, which speeds up the passage of nerve impulses. The relationship between primitive brain and higher brain is a fundamental part of human nature. Every day when survival is at stake, we call upon our fast-reacting animal brain to save our lives. The urban jungle in which most of us now live provides an environment just as rich as ancestral times for testing our primitive reflexes. Although our sophisticated cerebral cortex helps us to keep our wits about us, when we make a mistake it is our primitive brain that bails us out.

To put a human face on this cerebral coexistence, we can follow the adventures of a young man, a typical young man on a night out. In the course of a few hours he'll run a gamut of emotions – euphoria, fear, aggression, anger, frustration – and experience sexual arousal, all the while fuelled by one of humanity's oldest companions: alcohol. He may think he's in control of his actions, but primitive drives are pulling his strings.

However, this young man's night out almost doesn't happen. Let's say he's

been in a good mood all day, promoted at work and looking forward to celebrating in the evening. And his anticipation is heightened by the prospect of romance with a particular young woman he fancies. He's accompanying her across the road, when – DANGER! Speeding traffic bears down on them. His cerebral cortex has been distracted chatting up the woman: it is the primitive part of his nervous system that now springs into life.

The sound of screeching brakes sends sound waves to his eardrum. Tiny bones within the middle ear trigger nerve impulses that are quickly relayed deep into the brain. Millions of interconnected brain cells process the screeching sound and recognise danger. Some nerve impulses are directed to the part of the cerebral cortex, where sounds are analysed, while others flash down a primitive (in evolutionary terms) but automated part of the nervous system – the autonomic nervous system. Volleys of signals are sent down nerves all over the body. Adrenaline is let loose into the bloodstream from the adrenal glands that sit on top of each kidney.

His entire body becomes transformed to enable it to react to crisis. The fast-acting but subconscious nervous system makes his pupils dilate, his airways swell and draw in extra oxygen, his heart rate double and his hairs stand on end. In his gut, the walls of blood vessels constrict, decreasing its blood supply. (In an emergency, digestion becomes unnecessary and therefore ceases.) This primitive part of the nervous system has taken over his body to enable him to react quickly to the impending crisis and to return him and his companion to the safety of the pavement.

Our emotions are experienced and expressed in an area of the brain deep below the grey matter of the higher brain. The limbic system harbours our primal drives and controls emotions such as anger, sadness and lust. The physical expression of emotions, like our man's response to fear, is channelled to the rest of the body through the autonomic nervous system. This part of the nervous system also has a crucial role in controlling the moment-by-moment activity of all our internal organs. As an eminent French physiologist, Claude Bernard (1813–78), once put it, 'Nature thought it provident to remove these important phenomena from the caprice of an ignorant will'. This complex system, though, can sometimes overload our capacity for physical expression, and we react in an inappropriate way. One common example is when fear makes us lose control of our bowels or bladder.

This shows a hair follicle with testosterone at its base which enables it to grow vigorously.

## PASSING ON THE MESSAGE

At the very base of the brain (both anatomically and in evolutionary terms) sits the brain stem, through which all nerve impulses to and from the higher brain must pass to the spinal cord. The brain stem also contains its own nerve centres that control critical functions such as breathing, blood pressure and wakefulness. The crucial importance of the brain stem is reflected in the fact that it is the benchmark for death: when it stops functioning, brain death is said to have occurred. Various medical tests, performed to a very strict code of practice, can check whether an unconscious person on a ventilator has a functioning brain stem. If not, then with the consent of the patient's next of kin the ventilator can be switched off, in the knowledge that brain stem death equates to brain death which in turn equates to death of the patient.

From the brain stem descends the spinal cord within its own bony protection – the spinal column. Nerves to different parts of the body spread out from the spinal cord between each vertebral block. Discs of cartilage that separate one bony vertebra from another can sometimes slip and press on the spinal cord. A slipped disc is a common problem that causes pain and sometimes numbness and weakness in the area of the body from which the squashed nerve roots spread out.

Nerve impulses are electrical signals that pass along the length of a nerve until it reaches a junction with another nerve – a synapse. At such a point the electricity triggers the release of a chemical (neurotransmitter). This chemical then activates an electrical signal in the neighbouring nerve. When a nerve ends on a muscle, for example, the muscle can be instructed to contract or relax, depending on the type of nerve and the type of neurotransmitter. So almost as instantly as a thought is generated in our cerebral cortex, a series of nerve impulses can travel a couple of metres to stimulate a group of distant muscles in the foot.

## EVOLUTION OF THE MIND

During the nineteenth century, neuroscientists believed that the evolution of the human brain was reflected in the different stages of the developing foetal brain: passing from fish, to reptile, to bird and to lower and then higher mammals. Certainly the cerebral cortex becomes progressively larger with ever more intelligent species (although Neanderthal man had a much larger brain than modern man, it is unclear whether the cerebral cortex was any larger). Overall, however, different animals have different survival needs, so their brains have developed with different strengths to suit particular environments. While the immature human embryo may indeed bear a fleeting resemblance to a fish with gills, this does not justify the conclusion that the human brain once evolved from the brain of a fish.

It has been estimated that in order to be born with the full complement of 100 billion nerves, the brain of the developing human embryo must grow 250,000 new nerves every minute. But newborn infants are grossly unco-ordinated, hopelessly inarticulate and incontinent. The reason is that all these nerves are poorly connected and lack much of the insulating material that speeds up electrical impulses. The baby's brain could not grow any larger while in the womb, the skull size being limited by the size of the mother's birth canal. At birth, we actually have the same sized brains as chimpanzees, but then the human brain grows more quickly and for longer in relation to body weight. More white matter is laid down and, during the first two years of life, an incredible 30,000 new nerve connections are made each second.

The early lack of nerve connections allows newborn babies to demonstrate primitive reflexes that aid survival. For example, a sucking reflex can be induced when the corner of a baby's mouth is touched, or a grasp reflex when the palm of the hand is stroked. These reflexes are lost when neural connections to the cerebral cortex are made. However, they may recur in later life if dementia develops – due to the loss of neural control from the higher cortical centres.

## SEX DRIVE: THE ROLE OF HORMONES

The cerebral cortex influences all parts of our primitive brain including the hypothalamus, which is unique in that it is directly linked to a small but crucial gland which sits at the base of the brain: the pituitary gland. This is the master gland which controls the release of all hormones – chemicals secreted from a gland in one part of the body but having their effect elsewhere. The brain therefore has ultimate control of how our bodies behave, using both an intricate network of nerves and a careful mix of hormones. One of the body's most potent hormones is testosterone, a form of steroid. It is present in both men and women, but to such a greater extent in men that it is synonymous with maleness. Its effects are twofold: it not only stimulates male characteristics (androgenic effects), but also builds up non-sexual organs, such as muscle (anabolic effects). As ever, it is influencing our young man on his night out.

## The unkindest cut: the effects of castration

Long before the role of sex hormones was understood, societies had observed that the removal of testicles had a marked effect on males – whether animal or human. Domesticated animals bred for food were castrated to make them fatter and improve flavour – calves, for example, or cockerels. It was also noticed that castration made male animals such as stallions and dogs more docile.

In modern times, human males are castrated only for medical reasons or, in certain societies, as punishment. Throughout history, boys and men have been castrated for other reasons too. Perhaps the most obvious motive is reflected in the original meaning of 'eunuch' - from the Greek for 'one in charge of a bedchamber'. Naturally, castrated male attendants in harems would present no sexual threat to the concubines, and were widely employed in that capacity in the East for thousands of years. (However, this is not to say that sexual desire was always safely eliminated – see below.)

Another special reason for castration was the effect on the male voice. During puberty, testosterone, secreted from the maturing testicles, causes the male vocal cords to grow to almost twice the length of a female's. Consequently, vocal pitch in the male is lowered. If a boy's testicles are removed before puberty, then his vocal cords will be as short as a female high soprano's. But his chest and upper airways will continue to grow under the influence of growth hormone, potentially giving this 'castrato' voice a special power and timbre. This effect was exploited notably by the papal church choirs of the late sixteenth century: for about 250 years thousands of young boys aged eight to nine were castrated (using very primitive techniques), in the hope that their voices would develop the requisite quality for stardom in the opera houses.

The last major opera to feature a castrato was Meyerbeer's *Il Crociato* in Egitto, sung by Giovanni Velluti (1781–1861) on the London stage in 1829. However, the voice of a castrato can still be heard. In 1904, gramophone recordings were made of Alessandro Moreschi (1858–1922), the last castrato to sing in the Sistine Chapel.

Every 20 minutes, day and night, his primitive brain triggers the release of a wave of testosterone into the bloodstream. It is spread everywhere by the blood and it has effects all over the body. When testosterone reaches pores in the skin, for example, it triggers hair growth. The more testosterone there is in the blood, the more often men need to shave and the more hairs they have on their chest and other parts of the body. Paradoxically, hair on the head, especially at the temples, is prevented from growing by high levels of testosterone. Thinning of the hair is also age-related, so both males and females over 80 years old are commonly bald; but it is the relatively low levels of testosterone in young women that allows them to keep a full head of hair.

Other parts of the male body are also affected by the absence of testosterone. Eunuchs (see box, above) and men born without functioning testicles are taller than normal, as the growth plates at the ends of the long bones did not seal off at puberty. They have pale, hairless faces, a strong head of hair and a distribution of fat around the buttocks and hips, similar to women. Like many hormones, testosterone makes its greatest impact deep inside the primitive brain. Without it we have no sex drive – a characteristic that made eunuchs fairly reliable guards of Eastern harems. Many eunuchs, however, could achieve erections and took advantage of a plentiful supply of women and totally safe contraception. Infants and pre-pubertal boys are also able to have erections, suggesting that testosterone might be unnecessary. In fact testosterone is produced not only in the testicles; small amounts are also made in the adrenal glands and the prostate and from fat stores. Testosterone is unnecessary to achieve an erection, but vitally important in improving the sex drive.

Women are not influenced by their sex hormones, oestrogen and progesterone, in the same way that men are influenced by testosterone. This is best illustrated in the way that a woman's sexual drive is not normally affected by the changing levels of her hormones during the menstrual period. Furthermore, after the menopause – when the ovaries stop producing oestrogen and progesterone – she usually notices little change in sexual drive (though some women find it increases once they are free of the fear of pregnancy).

Common and distressing symptoms for a woman at menopause are hot flushes and mood swings, which can be reversed by replacement of oestrogen – hormone replacement therapy (HRT). Men do not usually have such an abrupt end to their testicular production of testosterone and have no such 'menopausal' symptoms. Those few men who do have a sudden reduction in testosterone production in later life (because of disease or accident) develop hot flushes and mood swings, just like menopausal women. This rare event could legitimately be called the 'male menopause'. However, this term seems to have entered modern jargon to describe the emotional reaction of ageing men, when they suddenly become aware of their gradual loss of sex drive and sexual allure (and develop a renewed interest in young blondes and racing cars!).

Libido generally declines with increasing years. Men over 60 usually have smaller testicles and consequently lower levels of testosterone than younger men. Nevertheless, some men over 80 years are still sexually active, while others are less capable or less inclined. The reasons are as much to do with the increasing likelihood of an underlying medical disorder, such as heart disease or diabetes and the drugs needed to treat them, as to do with lower testosterone levels or lack of sexual stimulation.

Meanwhile, our young man still has a full complement of testosterone coursing around his body – but he won't be relying only on his sex hormones to improve his chances in the mating game. Before meeting the woman he fancies, he's off home for a spot of grooming – which ironically may in fact negate any natural advantage.

## Increasing desire: aphrodisiacs

For thousands of years our ancestors have played at amateur endocrinology, trying to improve libido with the secretions of animal sexual organs. For example, around 1000 BC in India the Ayurveda system of medicine (inspired by ancient Hindu scriptures) recommended eating animal testicles as a treatment for impotence. Since then, increasingly sophisticated attempts have been made to reinvigorate males with flagging libidos using testicular extracts.

In 1889, at the age of 72 years, the French physiologist Brown-Sequard (1817–1894), injected himself with an extract derived from the testicles of dogs and guinea pigs. He reported an increase in physical strength and mental ability, as well as improvement of his digestion and a lengthening of the arc of his urine. Even though these effects were almost certainly due to the power of suggestion, Brown-Sequard's observations were on the right track.

At the turn of the century, a number of disreputable physicians made a fortune out of implanting monkey testicles (monkey glands) under the skin of men with flagging libidos. In 1932, efforts to isolate the active ingredient of testicles were rewarded with the discovery of testosterone. It was quickly found to be successful at improving the sex drive of both men and women. However, the benefits for women were usually outweighed by the virilising effects of testosterone, such as deepening of the voice and growth of facial hair. By the 1940s, the popularity for testosterone as a boost to sex drive started to wane, but at the same time it gained favour for another purpose – body building (see box, below).

## Pumping iron: body-building

The anabolic effect of testosterone – accelerating growth and repair of muscle tissue – did not escape the attention of body builders and competitive athletes for long. Derivatives of testosterone have now been synthesised to have a greater anabolic component compared with the androgenic. These so-called anabolic steroids have generated a multimillion pound market in their illicit use by athletes of all calibre. It is estimated that in the USA alone one million people abuse steroids, mainly for non-competitive body-building.

One final ironic twist played by the body on those musclebound macho men who take steroids is that it renders them infertile. This is because the hypothalamus reacts to high levels of testosterone by instructing the adjacent pituitary gland to stop sending signals to the testicles and so reduces the body's own production of testosterone and sperm.

Anabolic steroids could even be effective for use as male contraceptives, if it wasn't for the fact that men must be relied upon to take them!

The influence of a civilised society has had a powerful effect on our courting behaviour. Sexual advances have become far more subtle compared with the blatant signals given out by our evolutionary ancestors. The first step our modern man takes in his quest for a sexual partner is to make himself more attractive according to the fashions of the day. He shaves, combs his hair, covers his chin with aftershave, plasters on underarm deodorant and puts on clothes that enhance his male image. In many ways these manoeuvres disguise his sex drive: they wipe away the effects of testosterone. Shaving, for example, eliminates one of man's most overt signs of sexual maturity, although carefully groomed 'designer stubble' may redress this aspect. Many bald men (who must have high testosterone levels) wear toupees or comb great swathes of hair across their bare pates. The smell of

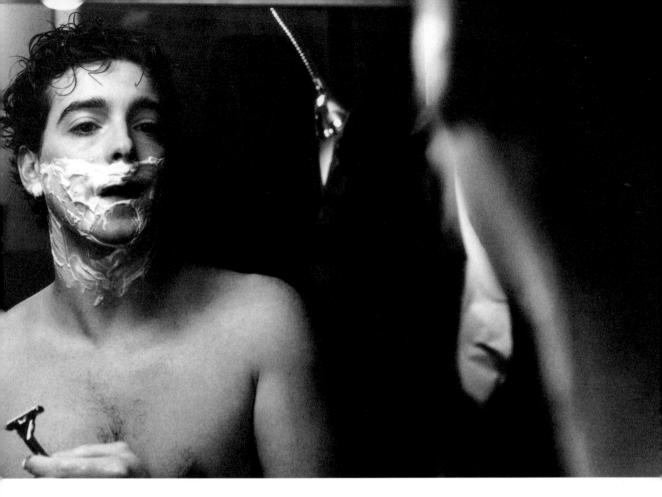

aftershave overwhelms the natural pheromones exuding from his skin and when slapped on in excess can be a turn-off for many women. The use of deodorant is another block on natural secretions originally designed to reflect sexual interest – see page 73. (Napoleon obviously knew what he liked when he sent a message to Josephine telling her to wait for him – and 'don't wash'!)

The style of clothes we wear are strong visual signs of our characters and play an important role in attracting or repelling a mate. At any time, the inverse logic that creates much of our modern sexual image can be replaced by primitive sexual urges released by the action of alcohol (which as the night wears on has more and more significance for our young man).

## THE DRIVE TO SURVIVE

The sex drive may be powerful, but it's more than matched by the basic instinct to survive in the first place. For this, we must eat food. The nervous system has several ways to make us eat: the stomach contracts and causes hunger pangs, while special

Testosterone is produced in the testicles but has far reaching effects. The amount of testosterone in a man's blood is directly related to how often he needs to shave, and gives him his sex drive.

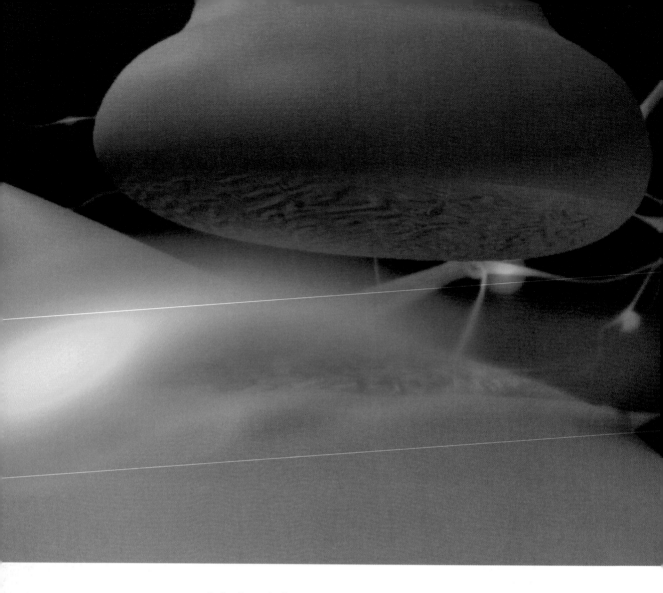

Dopamine is a neuro-transmitter, one of about 200 different chemicals that can bridge the gap between one nerve ending and another. This enables signals to move through the billions of nerves that make up the brain. Dopamine is released when we satisfy our survival pathways for food and sex. When dopamine is released our brains experience pleasure.

areas of the hypothalamus can sense low levels of glucose in blood. Even the thought of food can make us salivate – as demonstrated by Pavlov and his dogs.

To ensure our survival, the primitive brain also links the thought of food with the promise of pleasure. The act of chewing and swallowing triggers the release of a chemical called dopamine in the limbic system. Dopamine is a neurotransmitter – one of around 200 different chemicals that bridge the gaps between brain cells – enabling signals to travel from one cell to another and onwards through the brain. No one knows exactly how, but dopamine stimulates pleasure pathways in the brain. It gives a chemical 'high'. Our primary survival drives – for food, water and sex – are all inextricably linked to dopamine. Whenever we follow them, our brains release dopamine and we experience pleasure.

## SEX AND ALCOHOL

It is because our brains run on chemicals that we can fool them by using chemicals, hijacking the process behind our survival drives. Alcohol passes into the bloodstream from the stomach almost instantly, to be carried around the entire body. Alcohol slips with ease through blood vessel walls and into the surrounding tissue. Blood vessels in the brain are normally covered in a thick protective coat to prevent toxins entering – this is the blood–brain barrier. However, alcohol molecules are so tiny that they pass easily through the barrier; just seconds after being sipped, alcohol is floating through the fluid of the brain. Like every addictive drug, alcohol is a short-cut to a dopamine-induced 'high', acting directly on the limbic system.

### The need for dopamine

As we catch up with our young man on his night out – smartly dressed, closely shaved and doused in synthetic chemicals – he's in the pub with his mates, talking, laughing, bragging. He's been downing the beer and the result is that his brain is getting its chemical 'high' – but without him doing anything that will further his survival.

Tens of thousands of years ago males bonded before a hunt. Today only the terms of the hunt have changed.

The addictive nature of alcohol and other drugs such as nicotine and cocaine is related to the limbic system's ability to recognise these drugs and release dopamine. Sudden withdrawal of any of these drugs induces a fall in dopamine that is badly missed by the addict. The possible reason for some people having a more addictive personality than others appears to be due to a difference in the type of dopamine receptors in their brain. People born with less sensitive dopamine receptors may have chronically low levels of dopamine, which can be boosted only by drugs. If such people start smoking, then it becomes far more difficult for them to give up, as they return to their previously low dopamine state. Such individuals may benefit from anti-addiction treatment with a benign drug that stimulates the dopamine receptor.

**Sexual arousal**
The primitive brain is ceaseless in its quest for dopamine. In our young man's heightened state another survival drive – to reproduce – is kicking in. The limbic system is taking over his body in readiness for sex. His heart is racing, just as it was when he was threatened by traffic – but our emotional interpretation of a racing heart depends on the situation we're in. Now, he's not scared: he's aroused.

The blood vessels beneath the skin of his face are swelling, bringing more blood to the surface. By making us blush, the animal brain is showing our partner that we are sexually aroused. But our higher brain doesn't always want to reveal these feelings. The conflict between the two brains makes us anxious. Initially, our higher brain keeps us in check.

After two or three bottles of beer, the balance of power inside our man's skull shifts towards the animal brain. Like a sponge, his brain soaks up more and more alcohol. At the point where electrical signals pass from one brain cell to another, alcohol molecules dissolve into the cell membrane and start to block signals. The cell can no longer transmit messages onwards through the brain. Alcohol turns cells off all over the brain, but it has its first impact on the fragile cerebral cortex. Our social inhibitions are reduced by alcohol and our desire for sex is increased – as the porter in *Macbeth* puts it so pithily, alcohol 'provokes the desire, but it takes away the performance'. It reduces the ability to get an erection ('brewer's droop') and it also inhibits the ability to ejaculate. As the higher rational brain begins to slow and fade, the lower animal brain can take control. Thousands of years of evolution are dissolved in a few bottles of beer.

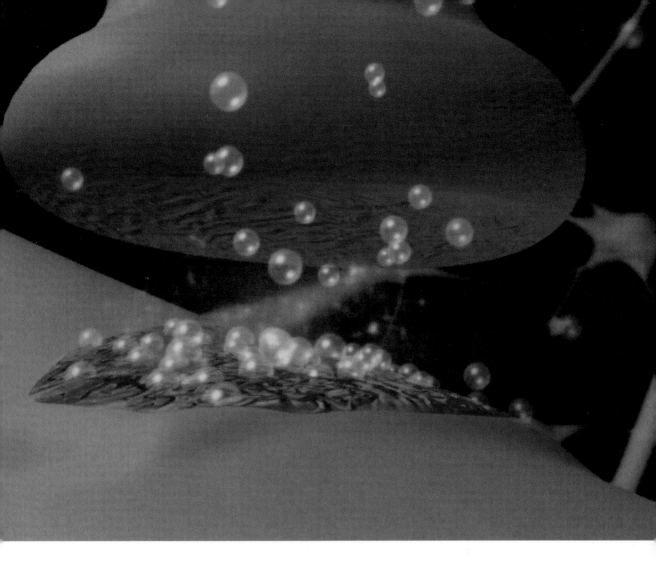

However, our young man isn't in such dire straits yet; he's had enough alcohol just to feel a pleasant loosening of inhibition. When the woman he fancies comes into the pub and his hopes of a sexual encounter improve, his primitive brain commands his body to produce sweat. Two and a half million sweat glands in the skin can make up to three litres of sweat in an hour. Although sweating is designed to cool the body down it has another, sexual, role. Deep inside the pores in his armpit, testosterone is being converted into a chemical called androstenol and carried by sweat to the surface. Bacteria living on the skin feast on androstenol, and the waste product of their meal is a strong scent. It may be the human equivalent of pheromones, powerful chemicals known to exist in other animals which trigger mating behaviour. Women who are ovulating are supposed to have a heightened ability to smell pheromones. When I was in medical school, a lecturer

**Above**: Alcohol molecules dissolve into the membrane of brain cells. They interfere with the passage of nerve impulses – turning brain cells off, one by one.

**Overleaf**: Not all brain cells are turned off by alcohol at the same rate. The cerebral cortex, responsible for analytical thought, is switched off first.

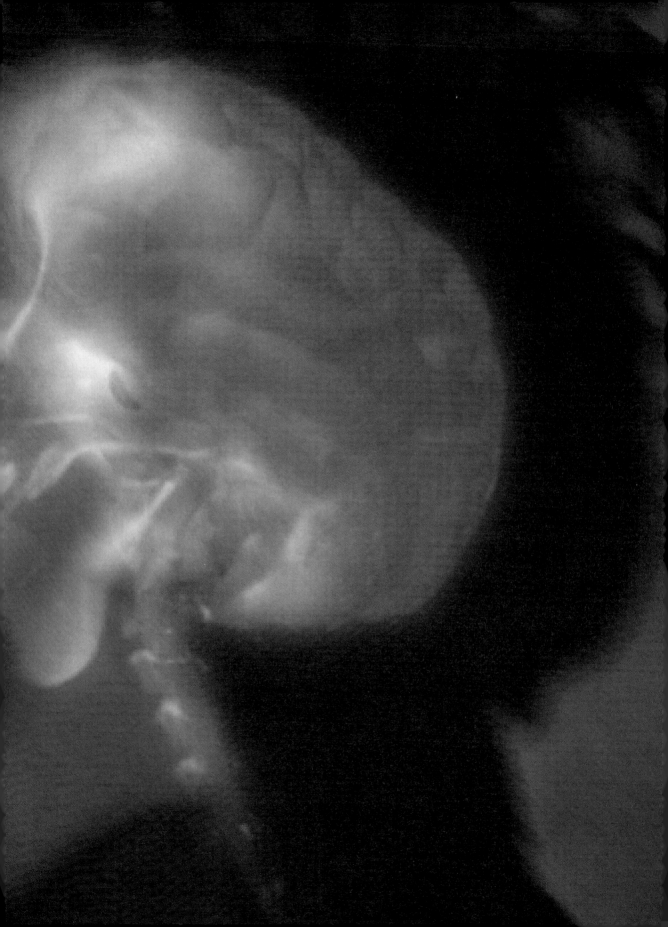

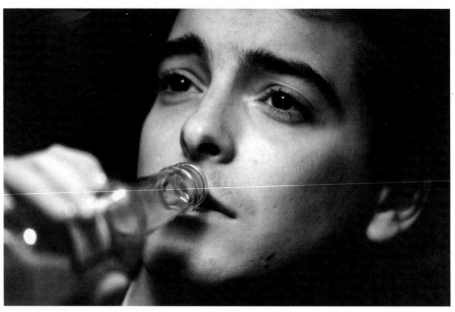

After just two bottles of beer alcohol starts to have a notable effect.

## An ancient comfort: alcohol

Our brains are no strangers to alcohol intoxication. Alcohol has been around for thousands of years, fermenting in rotten fruit and honey. Babylonian clay tablets from 6,000 years ago document a recipe for beer, complete with illustrations. Both wine and beer were widely drunk throughout western civilisations, providing a source of nourishment and hydration, in preference to the often dangerously contaminated water supplies. Westerners duly evolved a series of liver enzymes to metabolise alcohol and its breakdown product, acetaldehyde. Eastern populations were not so lucky: about 50 per cent of Chinese and Japanese people lack the enzyme acetaldehyde dehydrogenase. Just one glass of wine, taken by people without this enzyme, causes a very unpleasant reaction. Due to the build-up of acetaldehyde, they become flushed, dizzy and nauseated and collapse, without any of the pleasant symptoms of intoxication. A drug, disulfiram, has the same

effect by blocking this liver enzyme and so making alcohol unpalatable. It has been given to alcoholics trying to overcome their addiction; but of course it only works if the drug is taken.

Alcohol fermented from yeast has a maximum concentration of only 16 per cent – any higher would kill the yeast from which it was produced. Although this is strong enough to cause intoxication, its use as a food and a clean drink maintained its widespread consumption until the last century. Serious drinking of spirits did not start until the Arabs had mastered the art of distillation around AD 700. (The word 'alcohol' comes from the Arabic *al koh'l*.) They passed on this knowledge to Europeans but themselves chose abstinence. Excessive alcohol was soon recognised as a cause of ill-health (see box, page 83), but it wasn't until the end of the nineteenth century, when treated water became clean enough to drink safely, that alcohol lost its utilitarian function and was drunk for pleasure.

tried to demonstrate this point by passing around a vial that supposedly contained a solution of pheromones. Once all 100 of us students had sniffed this solution, he asked who could smell anything. He was hoping to identify only those female medical students who were ovulating, and about 15 out of 50 women put up their hands – but so did 20 men!

Of course, our young man has been blocking his body odour with aftershave and deodorant. But attractive perfumes can be synthesised from natural sources, taking advantage of natural odours – some perfumes utilise the scent from animal glands. Whatever fragrance is now adorning his companion, it has a powerful effect on our man. In the roof of his nostrils are nerves that detect smell. As scent molecules waft in they strike nerve endings, sending a signal straight into his brain.

The nerves responsible for detecting smell are the only parts of the brain exposed directly to the outside world. They can detect 10,000 different odours.

These nerves are the only part of the brain exposed to the outside world. They route smell directly into his limbic system, without going through the more highly evolved parts of his brain. This is why smell inflames unconscious animal drives, independent of our rational minds. Smells – and tastes – also have the power to evoke vivid memories of the past.

## Getting together

By now, our couple have left the pub and gone on to a party: finding a fairly private room, they step up their sexual activity. The man isn't the only one whose sex drive is inflamed by alcohol; the woman's level of testosterone is increased by alcohol too – with similar results. As they embrace and kiss, they are totally unaware that their primitive brains are talking to one another. Through the exchange of smell and saliva, they are giving each other secret information about their bodies. A man

Our lips and tongue are the most sensitive parts of our body. Kissing is the most economical way of stimulating the biggest part of the 'sensory area' in the brain.

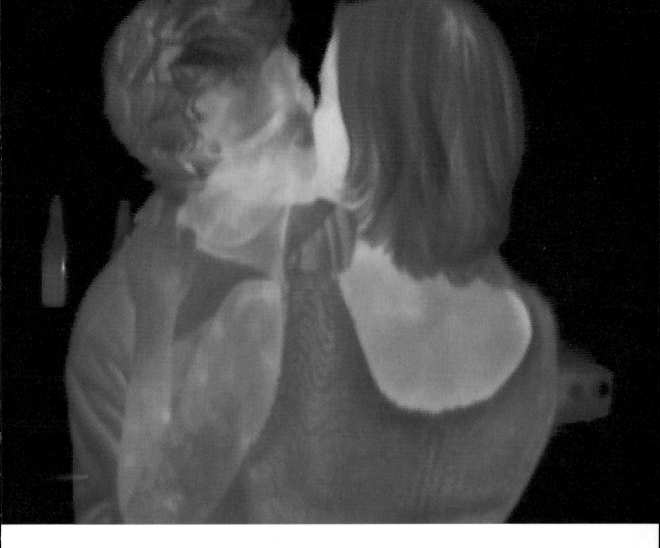

may be drawn to a woman for more than just the way she looks. He may also be attracted to her immune system. There is evidence that we become aroused by people whose immune systems are different from our own. Evolution may have designed us this way to ensure our offspring have a strong resistance to infection.

Of more immediate, and pleasurable, interest to our young man as he and the woman kiss is the physical sensation of touch. For this he can thank the fact that about 10 per cent of the area of the brain devoted to touch is linked to the lips and tongue. These areas of flesh were made this sensitive to help us articulate words, but it didn't take long for our amorous ancestors to work out that when we

Engorged with blood, skin temperature increases and the sense of touch is heightened.

## Flagging performance: problems of impotence

Failure to achieve an erection – impotence – is thought to afflict 20–30 million American men. It becomes increasingly common with age, often in association with cardiovascular disease, diabetes, depression, prostate surgery and the use of many different drugs. Until recently the treatment of impotence was unpalatable to most men and their partners. The use of injections into the penis or surgical prostheses has now given way to a drug, sildenafil (marketed under the name Viagra). In tablet form, it works within 30 to 60 minutes by blocking the breakdown of a molecule (cyclic GMP) that normally relaxes the penile arteries; in other words, cyclic GMP sticks around longer keeping the penile artery wide open. So far this new drug, which has been most widely used in the USA, has proved safe to use and therefore very popular. However, in the scramble to satisfy the patient's needs, doctors are tempted to overlook underlying medical causes for the impotence. There is currently no evidence that sildenafil enhances sexual performance in those who can achieve a normal erection.

are kissed, we are effectively having one-tenth of our bodies touched and stimulated in one go.

The biggest sex organ of any man is his brain. In its drive for pleasure, the primitive brain directs more and more blood to the surface of the skin and to the genitals. Flushed with blood, the skin temperature rises and, as it get hotter, touch receptors in the skin become more sensitive to a lover's kisses and caresses. At the same time, nerve impulses from the limbic system bombard the 'erection centre' in the lower portion of the spinal cord. From here, nerve impulses are directed towards the arteries of the penis, instructing them to relax. So much blood pours into the dilated arteries of the penile shaft that the veins leading out of the penis are squashed and blood cannot escape. This physiological response to sexual arousal is under the influence of the involuntary autonomic nervous system, but unusually it can also be blocked by instructions from the higher brain centres. It is this latter influence that inhibits anxious men from having a successful erection.

No such inhibition is affecting our young man. He has every hope that this sexual encounter will reach a satisfactory climax. Again, he will be unaware of the underlying mechanisms that work towards this goal. Just before male orgasm, the autonomic nervous system instructs the heart to beat more quickly, breathing to become rapid and perspiration to become more pronounced. Ejaculation occurs in two phases: emission of semen into the urethra – the tube shared with urine from the bladder – and then expulsion by muscular contraction. These two phases are stimulated by involuntary nerves between the genitals and the spinal cord. However, there are important influences on these nerves from the hypothalamus and limbic system within the brain. What finally triggers orgasm is not quite certain, but multiple influences include physical, emotional and hormonal factors.

### When things go wrong

However, our amorous young man is doomed to disappointment on this occasion. Just as things are warming up, an unwelcome interruption cools his ardour. He goes off to the bathroom and, in his fuddled state, manages to lock himself in. The previous emotions of the evening – euphoria, fear, arousal – give way to frustration and anger. His primitive brain takes more and more control of his actions. To think our way out of a crisis, we need our higher brain. If this is out of action, suppressed as in this particular instance by alcohol, our primitive brain is left to work out what to do. Its solutions are never subtle. Aggression rather than ingenuity is dominant, and our man hammers violently and fruitlessly on the bathroom door. However, sustained aggression demands energy. The adrenaline response is designed to supply it: adrenaline liberates stored sugar into the bloodstream to support physical activity.

Although alcohol blocks the inhibitory influence of the cerebral cortex on the limbic system, it is not always this pathway that leads to aggression. Testosterone undoubtedly makes an important contribution to aggression, as females are generally less aggressive than males. (An exception that proves the rule is the female spotted hyena, which is exposed to high levels of testosterone in the womb. The female has similar genitals to those of a male, with a greatly lengthened clitoris, and far more aggressive behaviour.) Furthermore, men convicted of violent crimes have been treated with anti-testosterone drugs and reported fewer aggressive and sexually deviant thoughts.

## Mass aggression

Throughout history, human society has usually been marked by displays of aggression. Whether fuelled by fear, hatred, greed, envy or religious divide, we seemingly need very little encouragement to take up arms against each other. At its most extreme, whole communities are persecuted and wiped out, in a concerted orgy of violence and cruelty recently christened 'ethnic cleansing'. And yet, to the puzzlement of many observers, these violent, cruel men (and they generally – though not always – are men) are also capable of leading quiet, respectable, law-abiding lives. One of the most notable examples of this apparent contradiction was the killing of European Jews by German soldiers during the Second World War. 'Ordinary' men, not noted for any tendency to aggression, were suddenly able to kill women and children with no compunction, and were then able to settle down with their own wives and children at supper. Less harrowing examples are seen nowadays soccer hooligans – usually 'tanked up' – whose violence terrifies innocent bystanders.

It has been suggested that this behaviour can be explained by a brain syndrome: Syndrome E. There is an area in the frontal lobes of the cerebral cortex which has evolved comparatively recently. When large numbers of people have been whipped up into an excited state, perhaps by a charismatic orator, this new area of the higher brain is hyper-stimulated. This leads to obsessive, repetitive acts of violence rewarded with feelings of elation. Similar states of high arousal and elation can be achieved using cocaine, which works in the same area of the brain and is highly addictive.

This stimulation of violence is different to that fuelled by excess alcohol, where the cerebral cortex is put out of action. In Syndrome E, appropriate physical or emotional responses to violence are absent, possibly because connections from the limbic system back to the frontal lobes are disrupted. The phenomenon is not seen in other animals (except possibly chimpanzees), making it unlikely that this behaviour results from the unleashing of the older brain. Ironically, what might be called 'animal' behaviour is due here to a development of the higher brain.

As alcohol travels further down the nervous system it sedates our cerebellum – so that we lose our sense of balance. The nerves that send pain signals back to the brain are also switched off. So when we fall over, drunk, we don't feel anything until the sedation has worn off.

It is still not clear where in the brain testosterone has its effect on sex drive and aggression. The most likely place is a small collection of nerves located just in front of the hypothalamus. These nerves appear to be two to three times bigger in males than in females, and twice as large in heterosexual men than in homosexuals. Exactly how these structures in the brain influence sexual orientation and sex drive (the latter is not diminished in homosexual men) is difficult to untangle. A purely biological influence on sexual behaviour and

## The down side: alcoholism

Over 15,000 years alcohol has seen varying degrees of popularity and decline, ebbing and flowing according to the dictates of the time. To take a recent example, the period between 1950 and 1980 saw alcohol consumption doubled while the relative cost of alcohol halved. Only about 10 per cent of the British population are teetotal. Medical advice in the UK recommends that men drink no more than 21 units of alcohol and women no more than 14 units per week. However, this level of alcohol is somewhat arbitrary and exceeded by more than 25 per cent of all men and at least 10 per cent of women.

It is estimated that each year alcohol contributes to about 100,000 premature deaths in the USA and 33,000 in the UK. Some of these deaths are due to liver failure, but far more follow from the effects of alcohol on the brain. Over time, the cumulative effect of excessive alcohol may lead to mental ill-health. Depression, anxiety, personality changes and sexual problems are linked to alcohol, which in turn may explain the association between alcohol and 80 per cent of all suicides, 50 per cent of all murders and 40 per cent of road traffic accidents.

There are three phases of alcoholic liver disease. The first occurs, to a greater or lesser degree, after every alcoholic binge – the liver becomes fatty. Nearly all of the liver enzymes are recruited to metabolise alcohol, leaving very few dedicated to the normal metabolism of fat, which therefore accumulates within the liver. This is a reversible state of affairs if drinking is stopped, but if alcohol abuse continues then liver cells are destroyed and alcoholic hepatitis develops. However, the liver has a remarkable capacity to repair itself, but in doing so the areas of regeneration disrupt the normal consistency of the liver. The formation of nodules of regeneration heralds the development of irreversible liver disease – known as cirrhosis. When this happens the liver is no longer able to function normally. Bile, a characteristic golden yellow colour, becomes trapped inside the liver and skin, leading to jaundice.

Only about 20 per cent of persistently heavy alcohol drinkers develop cirrhosis of the liver, although most will have a fatty liver. However, it is not just the amount of alcohol drunk, but an individual's susceptibility to alcohol that determines the risk of liver disease. This susceptibility is in part governed by the efficiency of the liver enzymes that metabolise alcohol. One study of men with alcoholic cirrhosis of the liver found that on average they had drunk the equivalent of two-thirds of a bottle of spirits every day for more than eight years.

Long-term use of alcohol has direct toxic effects on a man's nervous system. It reduces the production of testosterone, due to atrophy of the testicles. Furthermore, oestrogen levels are increased, leading to feminisation which causes swelling of the breasts and a female distribution of fat and hair.

aggression cannot be the whole story, as there are also strong influences from the society we live in. Drug dealers in New York City are more likely to find themselves in a violent situation than celibate monks in an abbey, despite both groups having the same testosterone levels.

Meanwhile, back in the bathroom, our man has found a hidden bottle of whisky and is drowning his sorrows even further. His anger and aggression are now diffused, and despondency is taking over. A kind soul releases him from his prison and he stumbles out into the night, cradling his bottle. In the past four hours, he has drunk a fifth of a pint of pure alcohol. His liver works so hard to break down the alcohol that other, more regular, liver functions may be disrupted, including the regulation of blood sugar. He starts to feel desperately hungry and gobbles up a takeaway pizza. An hour later, he has consumed another 10 units of alcohol. Now even his primitive brain is starting to shut down. This has drastic effects on the many organs which it normally controls. His kidneys draw too much

A hangover is caused by a dehydrated brain and the toxic by-products produced by the break down of alcohol. During the previous night, alcohol reduced activity to the area of the brain that stores memories. How we became in the state we find ourselves in the morning therefore becomes little more than a distant memory.

water from his blood, filling his bladder with fluid. However much fluid he drinks, his body passes out more. He's getting dehydrated.

At the rate he's drinking, alcohol is going into his body much faster than his body can get rid of it. Descending further, it reaches the base of his brain. Here, it begins to sedate his cerebellum, the part of the brain which helps us to keep our balance. Luckily for our man, the alcohol in his spinal cord has also switched off the nerves which feed pain signals into his brain. If it weren't for its dangerous side-effects, alcohol would make a perfect anaesthetic.

## The rhythm of life

Our bodies are driven by an internal clock situated deep inside the brain. The daily sleep-wake pattern is the most obvious example of our biological rhythms, synchronised to light-dark cycles and social time cues. Internal rhythms, also known as circadian rhythms (circadian from the Latin for 'about the day'), have major effects on our physical and mental performance. For example, our mood is usually lowest first thing in the morning, and our ability to carry out complex tasks hits a nadir in the middle of the night. The immune system also shows striking daily and seasonal variations. Depression is more common in the winter and, when it recurs every winter, has been given the fitting title 'seasonal affective disorder' (SAD). Suicides peak in the spring and autumn, ironically at the same times that the rate of conception is highest.

The network of signals that sets up the body's daily rhythm starts with light perceived at the back of the eye - the retina. (Blind people, who have no conscious perception of light, lack a daily biological cycle.) Information from the healthy retina is relayed along nerves to the hypothalamus, the part of the brain responsible for almost all circadian rhythms. Signals are then sent to the pineal gland, which sits in the very middle of the brain and produces melatonin – the darkness hormone. Every 24 hours at about 4.30 a.m., melatonin production peaks – up to 10 times its daytime levels. Melatonin is now available in tablet form and when given during daytime decreases the body's temperature and induces sleepiness. It has been of some benefit in minimising the symptoms of jetlag. Light and melatonin are the two most important elements that keep our biological clock in step.

Our man's brain is now totally saturated with alcohol. Barely any signals pass through his brain cells. At last his system can take no more. He crashes out in the street, dead to the world.

Next morning, our man begins to regain consciousness – though, tormented by the inevitable hangover, he probably wishes he hadn't. Pounding headache, dry mouth, nausea, dizziness ... his hangover has many causes. The water lost by his kidneys has left his brain dehydrated. His blood is full of toxic by-products produced by his liver's breakdown of alcohol. But enough alcohol has been cleared from his system to enable his higher brain to start functioning again – and it's just beginning to appreciate what his primitive brain drove him to last night.

As he groggily gets to his feet, our man can be thankful that he's not in charge of resetting his body functions. Once again the unconscious takes over and his basic instincts carry him through. The very rhythm of his life, his daily cycle awake and asleep, will be re-established without him even being aware of it. His body clock will tick over regularly – until the next night out.

# 4
# THE COLD WAR

Throughout our lives, we go to war against many ancient and invisible enemies: micro-organisms that must invade our bodies for their own survival. One battle we all fight is that against the common cold. It is a dirty war, provoked by an array of viruses using all their tricks to lay us low. Our throats rage, our noses stream, we cough, we sneeze, we shiver. Eventually we are turned into double agents, harbouring the enemy until it can safely infiltrate the allied camp: our neighbour. After a conflict that lasts several days, resistance builds against the invading virus and another battle is won – although the toll on our bodies means we need more time to recover fully.

**Previous page**: 2,000 antibodies per second are produced by special immune cells called B cells, in the lymph glands.

A number of viruses cause the common cold, but none is as deadly as the virus responsible for flu. The influenza virus is a different case, a real enemy, justifying all the big guns medical science can aim at it. Over thousands of years, the flu virus has taken on the human race in several world wars and won.

Most adults suffer about three 'colds' each year, while children have between four and eight (see below). Influenza is far less common, although it is reported that one in 10 of the world's population suffer it each year. I myself can remember having flu only once and will never forget it. I was in bed for a week, over Christmas 1996, and passed the infection on to my wife and two daughters, the younger of whom was only six months old at the time and far more ill than the rest of us. She didn't eat for 10 days and lost weight dramatically – almost back down to her birthweight. Thankfully, she made a full recovery. At the time I didn't even

## The common cold

Perhaps the eminent Canadian physician, Sir William Osler (1849–1919), had a point when he reputedly said that the only way to treat a cold is with contempt. Not that many other alternatives haven't been tried – the common cold causes more days off work or school than almost any other ailment. In an effort to stem the tide, the UK government opened 'the common cold unit' at Salisbury some years ago to find a cure. After thousands of volunteers were infected with cold viruses, it was closed down. Today we are little wiser about the best treatment of colds.

Two strategies can be used to treat the common cold. One is aimed at relieving symptoms and the other at attacking the virus itself. The most irritating symptom of a cold is a runny nose. Antihistamines are a group of drugs that reduce the volume of secretions from the nose and consequently the number of sneezes. One of the side effects of these drugs is drowsiness and therefore they are best given at bedtime, when an undisturbed night's sleep is valuable. Another group of drugs, anticholinergics, given as a spray up each nostril, can dry up the secretions and stop sneezing.

During a cold the blood vessels in the nose increase in size, causing a full, congested feeling. Some drugs stimulate blood vessels to constrict and therefore act as decongestants. However, these sympatho-mimetic drugs (so called because they mimic the actions of the sympathetic nervous system) act by narrowing arteries. This puts people with high blood pressure, and already constricted arteries, at risk of even higher blood pressure.

Steam inhalations improve the symptoms of a blocked nose in

some, but not others. It is harmless (as long as you don't knock the bowl of hot water into your lap!). Aspirin may reduce headaches, but has been blamed on increasing a runny nose, while decreasing the number of neutralising antibodies against the virus. Vitamin C appears to reduce the duration of a cold and the severity of symptoms by about 25 per cent. But large quantities (over one gram, equivalent to over 15 oranges) need to be taken each day.

Many people insist on being given antibiotics when they have a cold or flu, and many doctors oblige. But only bacteria, not viruses, are vulnerable to antibiotics. One survey found 60 per cent of patients seen by family doctors with the common cold were given antibiotics. It was estimated that the annual cost of antibiotic prescribing for the common cold in the USA in 1994 was about $40 million. This injudicious use of antibiotics is responsible for the developing problem of multi-drug resistance (see box, page 94). Occasionally bacteria take advantage of the damage caused by a virus and trigger their own infection, for example pneumonia. This is the time to use antibiotics in relation to a viral infection.

Antiviral drugs are also of no benefit against the common cold. Attempts to develop such drugs have failed because of the wide variety of viruses responsible. Furthermore, to be effective these drugs would need to be given before symptoms begin (the time that the virus is replicating most quickly); and those that have been tried cause nosebleeds and local irritation worse than the symptoms of the cold itself.

consider that my daughter might actually die from this bout of flu, but since then I have been left with little patience for people who use the terms 'flu' and 'cold' synonymously.

The name 'influenza' was coined in Italy during the Middle Ages, derived from *influentia*, the Latin for 'influence' — like a lot of afflictions, it was thought to be controlled by the stars. Nowadays, of course, the astrological theory has given way to the known fact that flu is caused by a virus — though it was not until this century that microscopes were powerful enough to look at them. Viruses are tiny: about 100 times smaller than bacteria and about 750 times smaller than a red blood cell.

A virus consists of little more than a packet of genetic material, enclosed within a membrane. It depends totally on another living cell to survive and replicate – outside a susceptible host cell it is inert. It's rather like a computer disk, which contains the code for a computer program but can be activated only when put into a computer. Although viruses carry their own genes, made of either deoxyribonucleic acid (DNA) or ribonucleic acid (RNA), they need to integrate themselves into a host cell's genes to replicate – and in order to invade the host, they use subterfuge. They are so successful that they have colonised almost every life-form on the planet, including animals, insects, plants and even bacteria. Viruses are the ultimate evolutionary organism.

While viruses are responsible for colds as well as flu (and indeed a whole host of other maladies (see A variety of viruses, right), there the similarity ends. The common cold is usually caused by the rhinovirus (from the Greek *rhis*, meaning nose), whereas flu is caused by a small family of influenza viruses, called A, B and C, which differ according to their protein make up.

Type A is the most common and, unlike types B and C, infects animals as well as humans. Type A is responsible for the massive flu epidemics that periodically affect the world. Its recurrent virulence is related to its ability to change its appearance and present a unique challenge to human and animal immune systems. Type B shares this ability to change its

## A variety of viruses

There are thousands of different viruses that have now been classified into different groups. Some are classified according to the diseases they cause, such as the herpes virus; some on what they look like, such as the coronavirus (surrounded by a 'corona', or halo of spikes); others on where they were discovered, such as the Marburg virus; and others on who discovered them, such as the Epstein-Barr virus.

Viruses are responsible for a wide variety of diseases apart from flu and the common cold, including hepatitis, AIDS, cervical cancer (papilloma virus), rabies, chickenpox and rubella. Indeed, it is this range of diseases, along with their small size and invisible mode of transmission, that has led to viruses being blamed for far more ailments than they are truly responsible for. A viral infection is a 'catch-all' for every time we feel off-colour, and very often a convenient label for a doctor to attach to a non-specific illness.

This image has not been helped by the continued uncertainty surrounding the cause of myalgic encephalitis (ME syndrome) or post-viral syndrome. It is still unclear whether this condition, characterised by fatigue after mild physical exercise and depression, is due to a past viral infection or is the physical manifestation of depression itself. The vague, but often debilitating, symptoms are certainly very similar to the physical symptoms of depression. In the absence of any solid evidence of a unique viral illness, it might be that in our society where mental illness is still stigmatised, viruses have again been blamed for a condition for which they are not responsible, and that some sufferers are not receiving the right treatment.

appearance, but much less dramatically than influenza A. It is responsible for the milder outbreaks of flu, which are especially common during the winter months. Type C influenza is not thought to cause serious ill health to humans, so this chapter will focus on types A and B which are a threat to us.

Influenza is probably the most infectious disease in the world. It is spread when an infected person (or animal) coughs or sneezes, expelling small droplets of mucus which are inhaled by the next victim. (A sneeze is a particularly efficient way of spreading viruses, propelled out of the lungs at about 100 miles per hour.) Every day we inhale 10,000 litres of air, and thousands of tiny airborne organisms – bacteria as well as viruses – enter our body. Other viral diseases may be spread in this airborne fashion or by other means (see How viruses are transmitted,

Just one of 100,000 globules of mucus expelled with a sneeze. This one contains the flu virus and a couple of much larger bacteria. The virus can only survive outside the body for a few minutes.

below). Where viruses even deadlier than influenza A, such as Ebola or HIV, are concerned, it's a mercy for us that they're not transmitted by the all too easy airborne route. The consequences would be catastrophic.

After a droplet infected with influenza A is inhaled, it can take up to six days for the symptoms of flu to develop – this is the incubation period. Influenza B takes only 24 hours before it makes its presence felt. Shortly after a person is infected, more than a million viruses can be found in each drop of nasal secretion. Although our immune system is usually effective at reducing this high number of viruses, we remain capable of infecting others for up to one or two weeks after catching flu.

Because flu is so infectious, sporadic cases are very rare. Usually it causes

## How viruses are transmitted

Although the flu virus is inhaled in an aerosol droplet and absorbed through the lining of the airways, other viruses invade in a variety of different ways. Some are swallowed and absorbed through the intestines, for example the polio virus – hence the polio vaccine is effective when swallowed on a sugar lump. Others infect through mucous membranes during sexual contact, for example herpes simplex virus type 2 and HIV; yet others through blood transfusions or shared needles, for example hepatitis B and C; while others infect through animal bites, for example rabies.

Once inside, the virus spreads directly to the host's organs and usually, but not always, kills the host's cells. An exception is the group of herpes virus, which lies dormant within nerves. When herpes is activated by sunlight, stress or other stimuli, a crop of blisters develop in the region of the infected nerve. Recurrent cold sores develop around the mouth when herpes simplex type 1 affects a nerve in the face, and genital herpes reappears when the nerves from the genitalia are infected with herpes simplex type 2. There is another type of herpes virus that is unusual because it is responsible for two diseases: chickenpox (also known as varicella) and shingles (also known as zoster, from the Greek for 'girdle' – the Norwegians call it 'a belt of roses from hell'!).

Chickenpox (no relation to smallpox), usually affects children and causes a typical rash of pustules, which contain the virus. Like influenza, this herpes virus is spread in droplets from the lungs or more rarely from the pustules –which are not very infectious. A person is infectious for 48 hours before the rash

develops and until all the pustules have crusted over. There is a lag of two to three weeks from the moment the virus first infects someone, until the rash develops. The herpes virus that causes chickenpox is similar to herpes simplex as it often lies dormant within a nerve –very rarely more than one nerve. Immunity against any further attack of chickenpox is usually lifelong but, having been infected, we remain vulnerable to shingles. Here, reactivation of the virus in later life causes a painful rash with pustules in the region of the infected nerve. People can catch chickenpox from the virus within the pustules of shingles, but shingles cannot be caught from someone with chickenpox or shingles –it is only a reactivation from the previous bout of chickenpox.

Two common viruses are particularly threatening to pregnant women and their foetuses. Chickenpox can be harmful to the foetus when caught in the first three months, and cause a severe infection in the newborn baby if caught from an infected mother within the last week of pregnancy. Rubella (also known as German measles as it was first described in Germany in the eighteenth century) can harm the foetus too, if caught by the mother during the first four months of pregnancy. The developing baby is vulnerable to congenital abnormalities, typically cataracts, heart defects and deafness. However, a vaccination programme against rubella has been very effective in dramatically reducing the numbers of congenitally acquired infections. Rubella is dangerous only during pregnancy; otherwise, it is mild, causing a rash in children.

epidemics, which vary in size from local outbreaks to global pandemics. (A disease becomes epidemic when it attacks many people in a region at the same time; a pandemic is a widespread epidemic, occasionally involving the whole world.) Here we can follow the progress of one victim of a small-scale epidemic, who had the misfortune to be in the confined space of a lift when someone sneezed. Her good fortune, however, was that the infected mucus she inhaled contained influenza B rather than A. If she had been infected with influenza A, her story could have been very different.

## A PATIENT'S PROGRESS

The body's first line of defence against airborne viruses such as influenza are the cells that line the nose and throat. They are tightly packed together and locked into position. The cells are coated in a thick protective substance known as keratin. On top of these cells are minute hairs (cilia) that project into the air passages. About 200 cilia on each cell beat 1,000 times every minute in organised waves, wafting away bacteria and viruses. (It's an advantage for our flu sufferer not to be a smoker – long-standing smokers damage these cilia and rely instead on coughing to get rid of invading micro-organisms. Despite this back-up system, smokers are still more prone to chest infections.) Interspersed among the other cells along the upper airways are goblet cells, so called because they secrete a film of mucus that coats the upper respiratory tracts. The mucus attracts cell debris, inhaled particles, micro-organisms and white blood cells. It is swept to the back of the throat by the cilia and either swallowed or spat out. The successful influenza virus must out-manoeuvre this first line of defence to seize control of the body.

Each virus is covered in little spikes that are used to stick into docking units (receptors) on the surface of the host cell. These haemagglutinin (H) spikes are so called because of their ability to glue red blood cells together (*haima* is Greek for blood, *gluten* is Latin for glue). They are not only crucial for the attachment and entry of viruses into the infected cell, but also determine its virulence. The surface of a flu virus is also covered with another type of spike, this time resembling the shape of a mushroom. These 'mushrooms' contain an enzyme called neuraminidase (N), which plays a key role in the release of viruses from infected cells. It is differences in the type of H spike (identified as H1, H2, H3 ... etc) and N mushroom (N1, N2, N3 ... etc) that distinguish one strain of influenza virus from another.

Once identified, a strain of influenza virus is given a name. Since 1980, a convention has been used that includes the place and year the virus was first recognised and the type of haemagglutinin and neuraminidase spikes. For example, so-called 'chicken flu', which threatened an epidemic in 1997 and infected millions of Chinese chickens, has been designated influenza A/Hong Kong/97/H5N1 (there's more about this outbreak on page 103). It is important to identify the exact type of flu virus that causes an epidemic so that a vaccine can rapidly be prepared against it. However, the speed at which a new strain of influenza A virus spreads makes it difficult to produce enough vaccine to protect any more than a few select groups. Our particular flu sufferer, being a young and

Different influenza viruses are distinguished by the spikes on their surface. The flu virus can change its spikes to present a new threat to people who have already had flu.

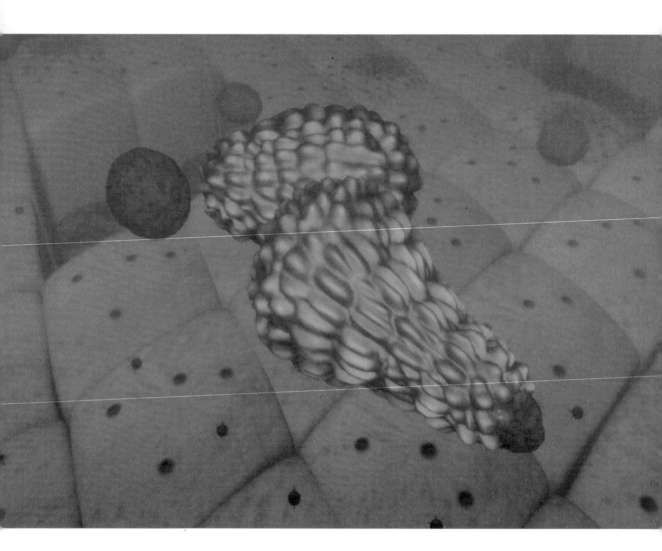

Macrophages are the body's clean-up crew. Here, two macrophages are eating pieces of infected throat cell that has been destroyed by natural killer cells. Macrophages release chemicals called cytokines that attract yet more macrophages to the throat and cause muscles all over the body to ache.

otherwise healthy woman, would not normally be considered for vaccination. Target groups include the medically vulnerable – the elderly and those with chronic ill-health, especially lung and heart disease – and people living in closed institutions where the rate of spread would be high. Important community workers who are valuable in times of an epidemic are also offered protection. The influenza vaccine is offered to the general public too but in the USA, for instance, it is estimated that usually only five per cent of the population ask to be vaccinated.

When the virus has locked into the docking unit on our patient's cell wall, it is taken inside. At this stage, it is disguised as a regular protein, wrapped in a bubble made from components of the host cell's wall. Once in, the virus is unwrapped and sent to the invaded cell's nucleus. Here, it commandeers the host cell's synthesising machinery and converts it into a dedicated cloning machine. From one virus invading one cell, 10,000 viruses are born. The invaded cells cannot stop the virus from repeating this feat of replication, but they can send out a smoke signal for help from the immune system.

One band of white blood cells are known as 'natural killer cells'; these roam the body on the look-out for trouble and are the first to respond to this smoke signal. They recognise infected cells and spray them with a harsh poison, killing the cell and the replicating virus. It's a clumsy response, involving the destruction of healthy cells too, but it's the immune system's only method of containing the spread of the virus at this stage. As a consequence of this battle, a lot of cellular debris accumulates, which must be disposed of to prevent choking.

The body's clean-up crew arrive. These scavengers, called macrophages, aren't killers – they eat debris. Just 24 hours after first being infected, our patient is aware of the battle raging in her throat as symptoms of flu begin. The carnage leads to the release of a chemical called histamine, designed to increase the flow of blood to the infected area, but the swollen blood vessels press on pain receptors, causing a sore

## Multi-drug resistance

Penicillin was the first antibiotic to be discovered, and was first used in 1940. Subsequent discoveries of other antibiotics (which are developed from micro-organisms) have helped enormously in the fight against infection over the last 60 years. However, at the time penicillin was being developed for human use, an enzyme known as beta lactamase was also discovered. This enzyme could destroy penicillin, and is responsible for bacteria becoming resistant to the antibiotic. It has in fact been around for thousands of years (discovered in antibiotic-resistant bacteria in the Canadian Arctic Circle), used by the organisms that produce penicillin, so that they don't kill themselves.

Although many bacteria remain persistently vulnerable to antibiotics, the list of bacteria and viruses resistant to antibiotic and antiviral treatment is growing. This is largely due to the overuse and misuse of antibiotics in human and veterinary medicine, as well as farming and plant culture. Antibiotics given to humans attack not only the sensitive invading pathogen, but the massive hoard of micro-organisms that grow in harmony within us. The human body is composed of 100 trillion cells, but only 10 per cent of these are human – the rest are bacteria, fungi, worms and even insects that make up our normal flora and fauna. These harmless organisms can develop antibiotic resistance and pass it on to invading pathogens. It is, however, the ability of bacteria and viruses to mutate that allows resistance to antimicrobial drugs to develop, and then in the company of such a drug the mutated strain has an advantage (over non-mutated organisms) and can replicate rapidly.

The answer to this problem is more judicious prescribing of antibiotics and antiviral drugs throughout human and veterinary medicine, as well as in farming practice. Good hygiene and vaccination programmes also have a role in reducing the need for antibiotic and antiviral drugs.

## Vaccination

Vaccination is an ingenious method of preparing an individual's immune system to fight infection. This is usually done by injecting the subject with a small amount of killed or inactivated micro-organism (e.g. virus or bacteria). The immune system recognises critical parts of the altered virus and sets up an immune response that gives just mild symptoms. The vaccinated subject is then immune to the more serious, often life-threatening, infection.

The first vaccination ever given was in 1798, against smallpox. Edward Jenner, an English country doctor, observed that milkmaids caught a mild form of 'pox' from cows but did not seem susceptible to the far more serious and often fatal infection, smallpox. Deducing that the lesser infection gave protection against the greater, he then vaccinated (taken from *vacca*, the Latin for cow) people by inoculating them with cowpox pustules. Recipients had only mild symptoms of cowpox, like the milkmaids, but were given immunity to infection from smallpox.

Smallpox is now eradicated in the world, though danger may remain from stored laboratory viruses (see box, page 104). These days, in the UK, everyone is recommended to receive protection against four viruses: mumps, measles and rubella (in a combination dose known as MMR), and polio. The aim is to eradicate these diseases, with vaccination given in childhood. The subsequent protection against these infections appears to be very good – probably lasting a lifetime as these viruses, unlike the influenza virus, do not change.

Natural immunity after infection with influenza depends on how much the virus changes, but can last for up to 10 years. Immunity following vaccination lasts for only one season and is specific to the strain that the World Health Organisation (WHO) predict will be prevalent that year. Being a smart outfit, the WHO usually hedges its bets and includes two influenza A strains and one B strain in a single vaccine. Even so, it sometimes gets it wrong. If the wrong strain of flu is anticipated, the flu vaccine is completely ineffective.

nearest to the site of the infection – the tonsils, adenoids and neck glands. Once within these glands, the cell meets other immune cells which can mount a highly targeted and focused attack on the influenza virus. These specialist cells, called T and B cells, are manufactured before birth and are so vital to survival that their total weight is more than the human brain. There are over a trillion of them, generated by our immune system as we have evolved. Each cell is designed to recognise and destroy a different invader. They spend their lives travelling around the body, from gland to gland, looking for the invader they are designed to kill – most will never find it.

Inside a neck gland, the immune cell containing the fragment of influenza virus is eventually recognised by a T cell. Suddenly the whole gland springs into action. The T cell that had been inert within the body since birth now clones itself hundreds and thousands of times. This profusion of cells causes the gland to swell and become tender. Our patient now has those familiar swollen glands in the neck and a raging sore throat, partly caused by swollen tonsils (which are glorified lymph nodes).

A single B cell within the gland also recognises the virus, and proliferates quickly – the immune system is on red alert. T-cell clones are launched into the bloodstream, and when they arrive at the battlefield they start a focused attack on the virus. Like natural killer cells, T cells destroy with poison, but they have a specific target firmly in their sights and they kill in much greater numbers. In a relentless march towards the lungs, the virus is attacking more and more throat cells. The hairy escalator that covers the throat cells and normally removes particles caught in the throat is temporarily out of order. The only way to get rid of excess debris is by coughing. The final assault on the virus by the immune system is about to begin.

Within the lymph glands the number of B cells that recognise the virus increases massively. Even when activated, the B cells never leave the lymph glands – instead they release protein missiles called antibodies at the rate of 2,000 a second. Antibodies are so small that they spread rapidly into all bodily fluids. They are also the most specific and versatile pieces in the immune system's armoury. Like heat-seeking missiles, they target the influenza virus and lock on to its spikes. The virus is rendered impotent: it can no longer clone itself and is now marked for destruction by macrophages. The combined forces of antibodies, T cells and macrophages begin to overwhelm the virus.

## T cells and HIV

The T cells that are vital in the fight against the flu virus are destroyed by the human immunodeficiency virus (HIV). This is usually caught during sexual intercourse or from contaminated blood. The immune system of a person infected with HIV becomes ineffective. Unchecked, roughly 10 billion new copies of HIV are made every day. Within two or three weeks after infection, a person develops a flu-like illness. At this time, the normal T cell count is cut by half. Within a week or so this acute illness is overcome, as other parts of the immune system strike back against the virus. But this is only a temporary reprieve; it seems that the virus is never completely eliminated.

During the chronic phase of the illness, people have antibodies against HIV (they are HIV positive), but are in good health. When the T cell count falls to about a fifth of normal (which may take anything from one to 20 years), the virus can no longer be held back and replicates rapidly. This is the stage at which people are said to have AIDS (acquired immune deficiency syndrome).

People with AIDS are vulnerable to 'opportunistic infections' – infections that a healthy immune system could easily deal with. Such infections include a rare type of pneumonia and even cancers, like Kaposi's sarcoma (caused by the ubiquitous herpes virus). Within about two years of these infections appearing, AIDS usually proves fatal. However, hope is on the way with the development of drugs that interfere with the virus's attempts to replicate within the T cell.

Until 1996, the death rate from AIDS in industrialised nations was climbing relentlessly, but now, thanks to various combinations of these antiviral drugs, it is starting to decline. This is little consolation to the millions of people with HIV or AIDS who live in developing countries, too poor to afford the drugs (one year's treatment costs over $10,000). It is a grim reflection of an unequal world that more than 90 per cent of people infected with HIV live in developing countries, but over 90 per cent of the money allocated for the management of HIV is spent in industrialised nations. But even in wealthy nations, there is no room for complacency. The ability of the virus to mutate and become resistant to drug treatment is already being seen, forcing new and multiple therapeutic strategies. In the developing world, drug treatment is unlikely to become widely available, but it is possible that the development of a vaccine might protect against HIV, although this is unlikely to be in widespread use before the early years of the next century.

If the invader had spread into the lungs, it could have caused a fatal pneumonia. More usually a pneumonia is caused by a secondary bacterial infection which takes advantage of the host's vulnerable state. This is especially the case in the elderly or in patients with pre-existing lung disease. It is for this reason that the elderly and physically vulnerable are the first to be offered vaccination against influenza (see box, page 98). Children with influenza are also vulnerable to a liver condition known as Reye's syndrome. This condition can be fatal, but thankfully is very rare. Aspirin has been implicated in causing Reye's syndrome and for this reason should not be given to children.

With the reduction of the invading virus, the biggest threat to the body is the growing army of immune cells. If the body does not recognise that the enemy is defeated, the immune system could overwhelm the body, like a cancer. Once the T cells have done their job, they turn their poison on themselves and the antibody-producing B cells, but not all T and B cells die. Some, known as memory cells, remain in the blood stream for ever. These memory cells make our flu sufferer immune to further infection by the same virus. If she is reinfected, the fight against the virus would be much quicker and more brutal.

Five days after our patient was infected, the number of viruses dwindles – she is effectively cured. She may not feel entirely well yet, as her immune response has taken its toll on her body, but she should feel completely well in another week or so. However, the flu story doesn't stop there ...

## THE FIGHT GOES ON

Our patient's immune system has defeated this attack by influenza B; the virus may well mutate and later present a new threat. But this is much more likely to happen with the A virus, which can mutate at an alarming rate, making it unrecognisable by those infected the previous season. It is this property of influenza A viruses that has allowed it to survive for thousands of years and to have such devastating effects on both human and animal life.

A new strain of influenza virus is born in one of two different ways. First, there is the mixing of viral genes, what is called 'antigenic shift'. If a cell is infected with two different strains of influenza A virus at the same time, their viral 'offspring' is created within the host nucleus; it is made up of a mixture of genes derived from both 'parent' strains of virus. If the changes affect the spikes that cover the virus, it can become unrecognisable by the human immune system and rapidly spread influenza throughout the population. This process often takes place between an influenza virus of human origin and one found within animals or birds. This century, there have been six epidemics of influenza occurring at irregular intervals: 1918–19, 1947, 1957, 1968, 1977 and 1989. In two of the three outbreaks for which viruses have been identified, the virus evolved by genetic mixing between a poultry and a human virus.

It also seems too much of a coincidence that three, if not all, of the influenza epidemics occurring this century originated in, or on the borders of, China (1957, 1968 and 1977). Why this should be the case is not completely understood, but one likely explanation is related to Chinese agricultural practice. Here, humans and domestic animals live in close proximity and the opportunity for two strains of influenza A to produce a new genetic strain is high. It is not clear why similar living conditions, say in parts of Africa, do not give rise to flu epidemics. One of the possible reasons may be that the virus prefers more temperate climates.

The second way that a new strain of virus may be born is called 'antigenic drift'. The existing genes of an influenza virus mutate at a very high rate, more than 10,000 times greater than that of human DNA. The changes in genetic material that occur because of mutations are less dramatic than those caused by gene mixing of two different strains of virus. They tend to cause local outbreaks of flu, rather than big epidemics.

### Overwhelming odds: great flu epidemics

Although ancient records tell us that disease epidemics have devastated populations for thousands of years, it has often been difficult to identify the responsible infection. One epidemic that remains a mystery but might have been flu was described by the Greek historian Thucydides at the time of the Peloponnesian War of 431–404 BC. This disease, which spread rapidly, proved fatal after a short and painful illness, killing up to a quarter of the Athenian population. Influenza was almost certainly around at that time, but so too were other viral diseases such as smallpox and measles, which proved equally fatal. The living conditions – where many people were in close proximity with huge herds of cattle – may have been particularly favourable to the flu virus, allowing it to perform the genetic shuffle that could out-manoeuvre the human immune system.

Humans are as bad as rats in their ability to carry disease. Whenever aggressive armies have invaded new lands, disease has ripped through the vanquished nations. Virgin immune systems are no match for virulent organisms. The spread of smallpox accompanied the rise of the Roman Empire, and the Spanish invasion of Central and South America in the fourteenth century. The fatal illnesses carried by invading armies weakened the ability of indigenous populations to put up resistance and made them easy to overcome. While infection may have been incidental to start with, it wasn't too long before invaders realised they could use this weapon deliberately, as a primitive form of germ warfare (see box, page 104).

A sore throat is caused by swollen blood vessels pressing on pain nerves.

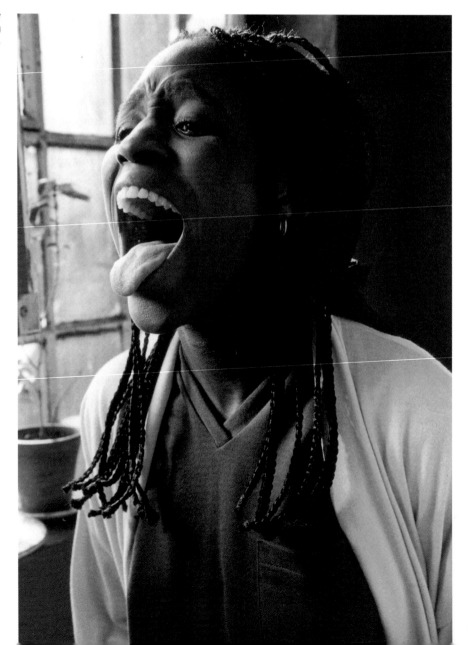

The first epidemic to hit the Americas appears to have been swine influenza, probably carried by pigs on board Columbus's ships. More than 400 years later, another outbreak of swine flu was to devastate the world. In 1918–19, a worldwide influenza pandemic followed the global movement of both service people and civilians after the First World War. More than 20 million people died (many more than had died in the war itself). It began in August 1918, and exploded at the same time in different cities around the world. By October, 20 per cent of the US army were ill and overall 24,000 soldiers lost their lives to flu, compared with 34,000 of their compatriots killed in battle. Its virulence was also apparent among civilians. In Boston, USA, 10 per cent of the population caught flu and 60 per cent of sufferers died. Little could be done to treat the sick or prevent the spread of flu. In many cities, public meetings were banned and the wearing of masks made compulsory. As quickly as it came, it vanished.

13 years later, in 1932, some basic research (always the cleverest) identified the virus responsible for this devastating pandemic. Scientists were able to transmit flu between pigs by taking brushings from the nose of an infected pig, and then smearing them on to the nose of another. Shortly afterwards, the influenza virus was identified. Two years later, it was shown that blood taken from people exposed to the 1918–19 flu pandemic had antibodies to the 'swine flu', but children born after 1920 had no such antibodies. A flu virus had developed as a hybrid between a pig virus (the haemagglutinin spike) and a human virus (the neuraminidase component). It was designated H1N1 and probably began in Asia, despite frequently being referred to as 'Spanish flu'.

A recurrence of this strain of flu was suspected in January 1976 at Fort Dix, USA, when a young army recruit collapsed and died after a training run. By the end of the month, around 300 of his colleagues had either become ill with flu or were confined to quarters. Analysis of brushings from the throats of these soldiers revealed the H1N1 virus, indistinguishable from the flu virus responsible for the 1918 pandemic. The question on everyone's mind was whether this was the beginning of another worldwide flu epidemic. This was not an easy question to answer, as many of the army recruits at Fort Dix were also infected with another influenza A virus prevalent at the time. Nobody could be certain about the infectivity and virulence of this pocket of 'swine flu' at Fort Dix. Controversy reigned among those advising the government on whether or not to initiate a massive vaccination policy. Eventually, it was decided that the consequences of not

## Pandora's box: researching for war

There have been well-documented cases of lethal viruses escaping from laboratories and taking lives. One of the most notable was in 1978, two years after the successful worldwide eradication of smallpox. A British photographer was working above a laboratory at the University of Birmingham, UK, from which the smallpox virus escaped into the ventilation system. She caught the disease and died.

At present only two laboratories have official stocks of the smallpox virus, in Novosibirsk, Russia, and the Center for Disease Control in Atlanta, USA. It is a source of great anxiety to imagine which governments around the world have held on to illegal stocks of smallpox and other infective agents as ammunition for biological warfare. Indeed, smallpox was the first virus used in biological warfare, in 1763.

At that time, Fort Pitt on the Ohio River, USA, was not only under siege by tribes of Indians but was suffering the ravages of an outbreak of smallpox. Colonel Bouquet, the commander of the fort, arranged for blankets previously used by those suffering from smallpox to fall into the Indians' hands. The Indians became infected and many thousands lost their lives, as did many of the invading colonists.

A dentritic cell containing a fragment of virus locks into a T cell. This T cell will then replicate thousands of times and start a focused attack against the virus.

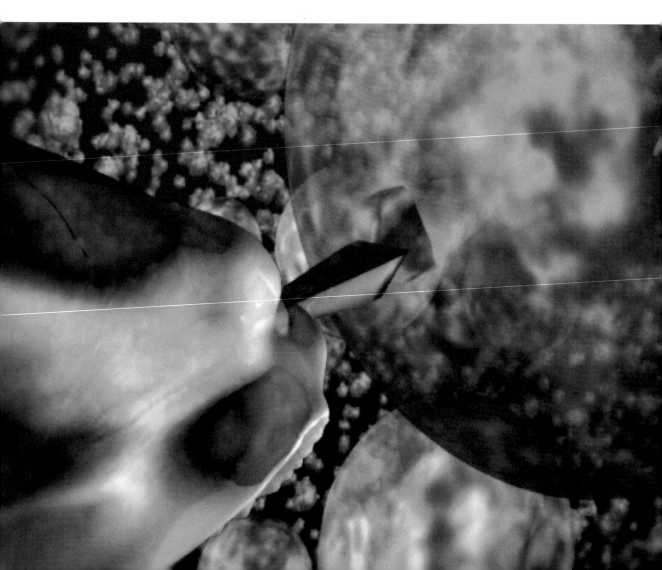

embarking on a mass vaccination programme, if this was truly a resurgence of the virulent 1918 virus, would be so catastrophic that no risk should be taken.

Millions of dollars were spent on producing gallons of the vaccine and about 80 per cent of the US population were vaccinated. In the end no flu outbreak ever emerged. After nearly 60 years among pigs, it is possible that the H1N1 influenza A virus had changed its genetic make up into one of low virulence. Or maybe poor living conditions and mass population movement at the end of the First World War combined to create the right melting pot for a flu virus that the people of the late twentieth century could take in their stride.

Historical records suggest that another influenza pandemic is overdue. The next will occur when an influenza virus emerges to which most people in the world have no immunity. The threat of this happening is always with us, because of the chameleon-like ability of the flu virus to change its identity and evade a host's immune system.

As with the Fort Dix experience, the threat of a new epidemic certainly concentrates the medical mind. In Hong Kong, in August 1997, a child died from liver disease and an autopsy revealed the presence of 'chicken flu'. Although this type of flu was highly infectious and lethal to chickens, once a flock was infected the mortality rate was 70–100 per cent, it rarely infected humans. One research study showed, for example, that it infected only five of 29 poultry workers and only one of the 54 health-care workers who looked after the first human case. Nevertheless, it was felt that the only way to prevent a flu epidemic in humans was to frustrate the attempts of a highly virulent strain of flu to emerge from poultry. Millions of chickens in Hong Kong were slaughtered and, thankfully, no further human cases of influenza A/Hong Kong/97/H5N1 have been identified. Whether or not 'chicken flu' would have led to a human epidemic, even if the mass slaughter of chickens had not taken place, remains unclear, but the evidence from this outbreak suggests this was a smart move to outwit the virus.

## Fighting the enemy

Research work on the highly infectious influenza A virus can only be carried out in specialised laboratories that are sure to contain the spread of the disease. This is a great challenge, as the flu virus is so easily transmitted in air droplets. Such laboratories are therefore designated a higher safety level than those in which the far less infectious, but more lethal, HIV (human immunodeficiency virus) is

**Overleaf:** Two members of the body's immune system are T and B cells. They specifically recognise the virus and start a relentless attack to wipe it out.

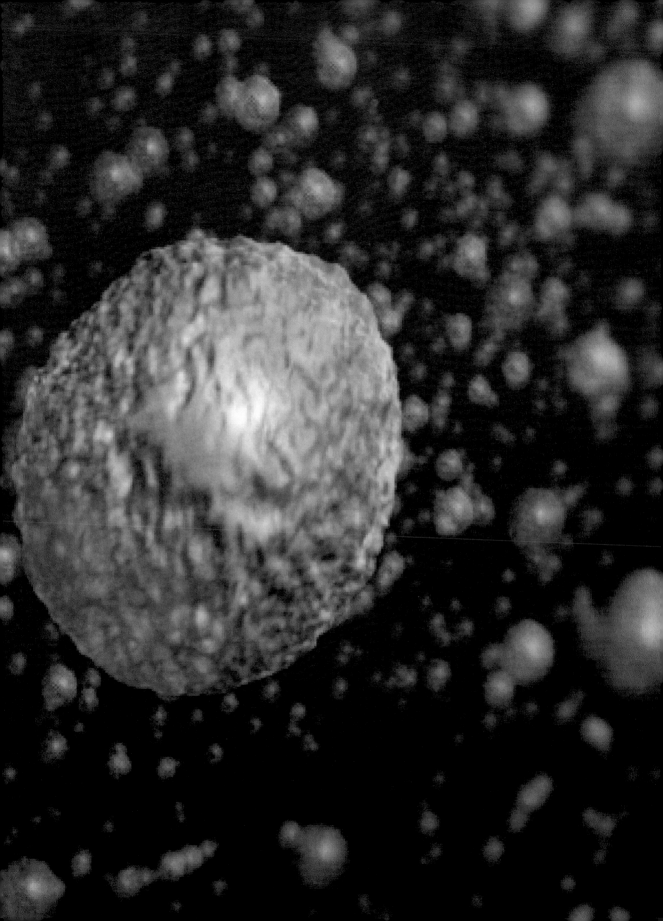

investigated. At present very few of these highly safe laboratories exist, but the standards needed to contain these viruses must never be allowed to slacken. There was a scare in 1976 when a strain of flu virus was identified, almost identical to the virus responsible for the 1947 influenza pandemic. This could well have been an escape from a research laboratory, although it might also have been due to remarkable genetic stability in some non-human host.

There are drugs to treat viruses, but as yet none is really useful in the treatment of influenza. One of the main difficulties in developing antiviral drugs in general is that they must kill the virus without damaging the host. As a virus spends a large part of its life within the nucleus of a host cell, this is a difficult proposition. However, there is a drug effective against influenza A, known as amantadine. Although it is not precisely known how it works, it is thought to interfere with the virus's attempt to unwrap itself after being taken inside the cell; the genetic material of the virus is therefore unable to get to the host cell's nucleus. The limitation of amantadine is that the drug needs to be given during the first day of illness, while the virus is still invading the host.

Antiviral drugs are not only poisonous to healthy cells, they quickly become ineffective as the virus mutates and becomes resistant to the treatment. Under the circumstances, the policy of antiviral vaccination looks a better prospect. However, this is not to dismiss antiviral drugs altogether – they are vital in the fight against other infections. In particular, a great deal has been learnt about them in the treatment of HIV (see box, page 99).

## TURNING THE ENEMY: GENE THERAPY

So far, we have been concerned with the battlefield where human host and invading virus struggle for supremacy. Medical science has always aimed at the death of the enemy. Now, however, in an ironic twist, viruses may be requisitioned for the good of our health.

The uniqueness of every living thing is determined by its genes. However, some of our important genes govern essential biological functions, and are shared by everyone. People born with a critical gene missing, or non-functional, might die or become ill: they have a genetic disease. In recent years much has been learnt about our genetic make up and the abnormal genes responsible for certain diseases. Over 4,000 diseases are now recognised to be caused by abnormal genes.

Medical science has developed techniques of isolating and manipulating healthy genes in order to transfer them into a diseased cell: this is gene therapy. It is a new field and there are still many major difficulties to overcome before it is widely practised. One of the greatest problems is delivering the healthy gene to the abnormal cell. Genes sit on chromosomes deep within the nucleus of a cell and are therefore very difficult to access. One ingenious solution to this problem is to take advantage of the natural ability of viruses to integrate into a host cell's genetic make up.

A virus can be manipulated so as to carry a healthy gene instead of the gene that allows it to replicate. When this altered virus infects a target cell, it activates its own genes as well as the healthy gene that the host cell lacks. At the beginning of 1997 just over 2,000 patients worldwide had been treated in this way and numerous gene therapy companies have emerged, trying to master the technique. The disease most often treated is cancer, using viruses that carry genes encoding for lethal proteins, specific to the tumour cells. Some patients with cystic fibrosis have also tried gene therapy. In this case an aerosol inhaler delivers a healthy cystic fibrosis gene attached to an adenovirus (one of the viruses that normally causes the common cold) into their damaged lungs. So far, however, this treatment has proved disappointing as many of the cells do not activate the healthy gene. Furthermore, genes delivered to a patient who already has damaged lungs might prove to be of little benefit.

This has prompted some investigators to look at the possibility of delivering healthy genes to foetuses that are known to have inherited a gene defect. This work is still experimental and has not yet been performed in humans. It is also worth noting that gene therapy has nothing to do with the modification of genes within sperm or eggs before conception (an illegal practice in the UK).

In this chapter we have seen the different strategies used by both viruses and humans in their long-standing battle. The flu virus is one of the most cunning and deadly. Periodically it re-emerges in a different guise, outwitting the human immune system and killing millions. The development of flu vaccine tailored to the flu virus is a masterful counter-attack – so at times both sides could claim a breakthrough. But at present the innate versatility of the virus and the endless ingenuity of humans have reached an unstable impasse.

# 5

# UNDER

# PRESSURE

Most people believe that the heart is the most important organ in the body. It certainly does the most dramatic job and has the most dramatic impact when it fails. The more sophisticated organs, such as the brain, liver and kidneys, rely on the heart for delivery of their fuel — oxygen-rich blood.

**Previous page:** Red blood cells circulate around the body delivering oxygen to distant tissues. They then return to the heart and lungs for a fresh pick up of oxygen. On average red blood cells can continue this relentless task for four months before being replaced.

The human body is a remarkable society of around 100 trillion cells. Each one of these cells is the direct evolutionary descendant of a tiny, single-celled organism which once moved about to find food. However, within the human body most of our cells live in a fixed position. They cannot scurry about to find food; instead, food must be delivered to them. A network of blood vessels more than 120,000 kilometres long – equivalent to three trips around the world – transports oxygen-rich red blood cells to their point of delivery. These red blood cells squeeze through ever narrower blood vessels so that eventually every fixed cell in the human body lies within 20 micrometres of a blood vessel – a distance one-third the thickness of a human hair. At this point, red blood cells release their oxygen and take on-board carbon dioxide, the waste product of cell metabolism. Without a discernible pause for this exchange to take place, the cells are pushed onwards to begin their return journey to the source of this ceaseless motion, the heart.

The heart is almost pure muscle. It weighs about 300 grams, beats around 70 times a minute and pumps more than 4.5 litres of blood around the body every minute of every day. It has been designed as two pairs of chambers and four valves.

There are two collecting chambers, the atria, and two pumping chambers, the ventricles. The valves control the flow of blood out of each of the four chambers. Blood returning from the body, low in oxygen, enters the right atrium. On every beat of the heart a certain amount of blood, equivalent to about half a cup of tea, has collected; then the valve into the right ventricle opens, the blood falls in and the valve shuts behind. The valve leading into the pulmonary artery then opens, and blood is pumped out of the right ventricle and around the lungs to pick up oxygen.

The sound of the heartbeat – the familiar 'lub-dub' – is made by the valves being opened and closed; it can be heard very clearly through a stethoscope. Like a central heating system, abnormal sounds are caused by a narrow or leaky valve or when the heart muscle itself starts to fail.

Blood propelled around the lungs not only picks up oxygen but also releases carbon dioxide. To satisfy the demands of the body, huge volumes of these gases need to be exchanged within a fraction of a second. This is made possible by two extraordinary feats of bio-engineering. Firstly, the wall separating blood from

## Replacement valves

A badly diseased heart valve can be replaced by one made from human, pig or cow heart tissue, or by a mechanical one. Since the 1950s more than 80 models of prosthetic heart valve have been developed and more than 60,000 heart-valve replacements are performed annually in the USA alone. Mechanical heart valves last about 20 to 30 years and are considerably more durable than those made from heart tissue. However, they do have some drawbacks that do not apply to the natural tissue. Blood tends to clot on mechanical heart valves, so patients fitted with them must stay on blood-thinning treatment for the rest of their lives. Also, one type of artificial valve consists of a ball that flips up and down within a metal cage. The ball clicks loudly on opening and closing, making a noise that can clearly be heard without a stethoscope – an irritating problem for the patient (and anybody close by).

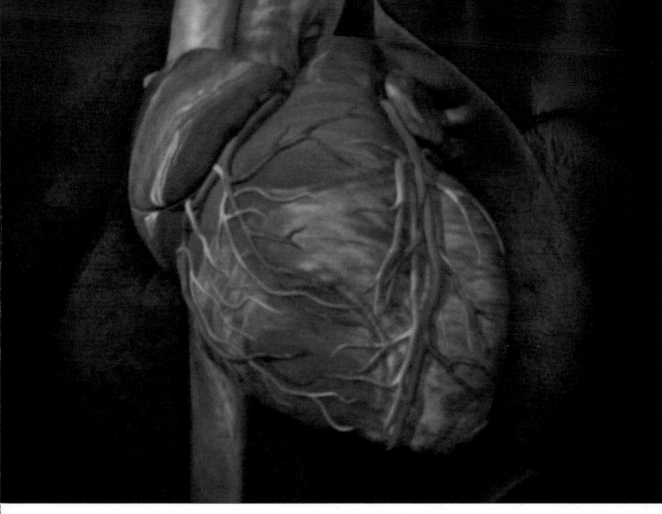

## The developing heart

The heart begins to beat only 23 days after conception. At this time, it is still only a single tube within the developing foetus. Its transformation into four pumping chambers requires a series of folds and twists of extraordinary complexity. Furthermore, at the moment of the first breath, the baby's heart undergoes even more momentous changes.

Inside the womb, the developing baby receives all its oxygen from the mother via the umbilical vein. All its waste products are delivered back to the mother through the umbilical artery. The foetus has no use for its lungs. Blood is therefore bypassed directly from the right to left side of the heart through a communicating hole and bypass duct. At the moment of birth, the baby is suddenly on its own and must breathe for itself. Blood is immediately diverted into the blood vessels of its lungs to pick up oxygen. Both the hole in the heart and bypass duct, through which blood was diverted, must be shut off. If they aren't, blood continues to bypass the lungs and the baby becomes blue from lack of oxygen. Such a baby has a persistent 'hole in the heart'. Nowadays, operations to close off holes in the heart have become routine and these children do very well.

Perhaps, considering the complexities involved in hearts, it is a wonder that more babies are not born with congenital heart defects.

The healthy heart muscle showing the two main coronary arteries which lie on the surface and the many branches off these arteries which feed the heart muscle with oxygenated blood. A narrowing in any part of this coronary artery system reduces blood supply to the heart muscle, causing angina. A complete blockage causes a heart attack.

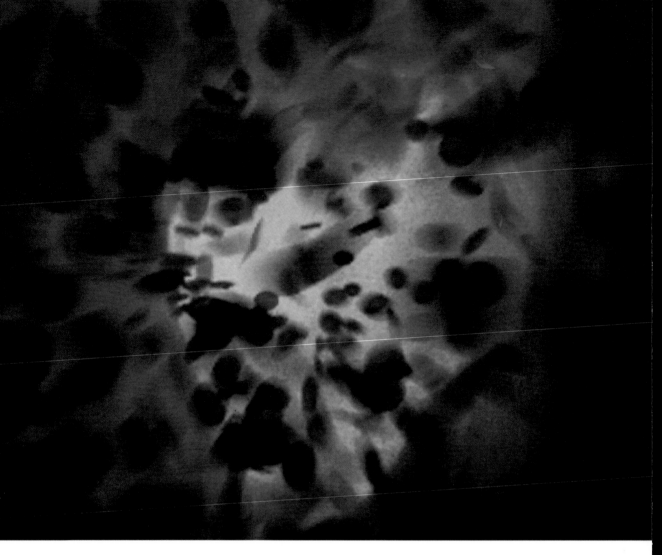

air within the lungs is vanishingly thin: little more than 1 micrometre – 50 times thinner than a sheet of airmail paper. This explains why patients with chest infections or pulmonary tuberculosis, their thin lung walls eroded away, readily cough up blood. Secondly, there is a large surface area for air to come in close contact with blood. Laid out flat, this would be equivalent to the size of a tennis court. Charged up with oxygen, blood then enters the left atrium through the pulmonary veins – the only veins in the adult body that transmit arterial (oxygenated) blood. From here it goes to the left ventricle which pumps blood into the biggest blood vessel within the body, the aorta. Numerous arteries leave the aorta to deliver blood to the body's tissues, through smaller and smaller blood vessels. Finally, blood reaches close enough to cells fixed in organs remote from the heart, for oxygen and other nutrients to be unloaded as fuel for cell metabolism.

In order to beat continuously for every moment of our lives, heart muscle – unlike any other muscle in the body – has a life of its own. All of its cells have the ability to beat independently, but normally they beat in unison. Each heartbeat is co-ordinated by clusters of special cells buried deep within the muscular walls of the heart. These cells act as a natural pacemaker. They send electrical impulses throughout the heart to faultlessly co-ordinate the speed and force of every beat.

The tireless action of the heart necessitates a good blood supply of its own. This is supplied by two coronary arteries, the right and the left. Both these arteries arise from the root of the aorta, just beyond the left ventricle itself. They course their way on the surface of the heart with branches that spread into the heart muscle (called the myocardium, from the Greek for 'muscle heart'). For reasons that will be discussed later, the coronary arteries are particularly vulnerable to a disease that narrows them – atherosclerosis. Derived from the Greek word for 'gruel', atheroma can develop in any artery. Formed by a combination of blood vessel wall damage and excess free-floating fats, the build-up of atheroma leads to narrowing of the artery. Downstream of a narrowed artery, the reduced blood flow leads to a condition called ischaemia – when cells become deprived of oxygen and food. Within the coronary arteries, atheroma leads to coronary/ischaemic heart disease or, if the vessel is completely blocked, a heart attack – myocardial infarction.

High blood pressure or turbulent blood flow can damage the inner lining of a blood vessel. In response to the injury, white blood cells, which usually fight infection, adhere to the damaged inner lining. They then edge their way through the vessel wall, releasing chemicals to attract other circulating cells. Among these cells are platelets, sticky components of blood, which clot together over the surface of the injured vessel. In the meantime the underlying white cells have eaten the excess fat and become stuck in the vessel wall. The muscle cells in the wall of a blood vessel, which normally allow it to constrict or dilate, also increase in number. All these different components build up over time to form the main sign of atherosclerosis, the 'plaque'. Paradoxically, it is

## Missing a beat

Occasionally, the heart's own pacemaker system fails. Impulses from the atria cannot get through to stimulate contraction of the ventricles. This is called 'complete heart block' and occurs in the elderly or following a heart attack. Patients with this problem may not have an effective heartbeat for periods lasting a minute or more. They collapse in a faint, without a pulse, and look as if they are dead. Life is restored when a heartbeat resumes and blood is again pumped to the brain. That a complete heart block does not lead to sudden death on every occasion is thanks to a remarkable back-up system. Whichever level of the heart's conduction system is blocked, a new natural pacemaker, below the level of the block, can take over and generate nerve impulses. However, instead of beating 70 times a minute, the cells in a ventricle that have lost any influence from the natural pacemaker beat around 40 times a minute. The problem is that contraction of atria and ventricles is no longer co-ordinated, so both sets of chambers beat independently and often against each other. The treatment is to insert an artificial pacemaker. These devices, no bigger than a cigarette lighter, are placed under the skin in the chest. The most basic pacemakers have a wire that leads into the right ventricle and stimulates muscle contraction 70 times per minute. The only trouble is that pacemakers run on batteries and batteries run out, so they need to be replaced regularly.

**Overleaf:** Here the red blood cells have reached the end of the arterial tree and oxygen is released to fuel the metabolism of nearby tissues. There are about five billion red blood cells in the human body. They are biconcave disks, about seven micrometres in diameter with no nucleus.

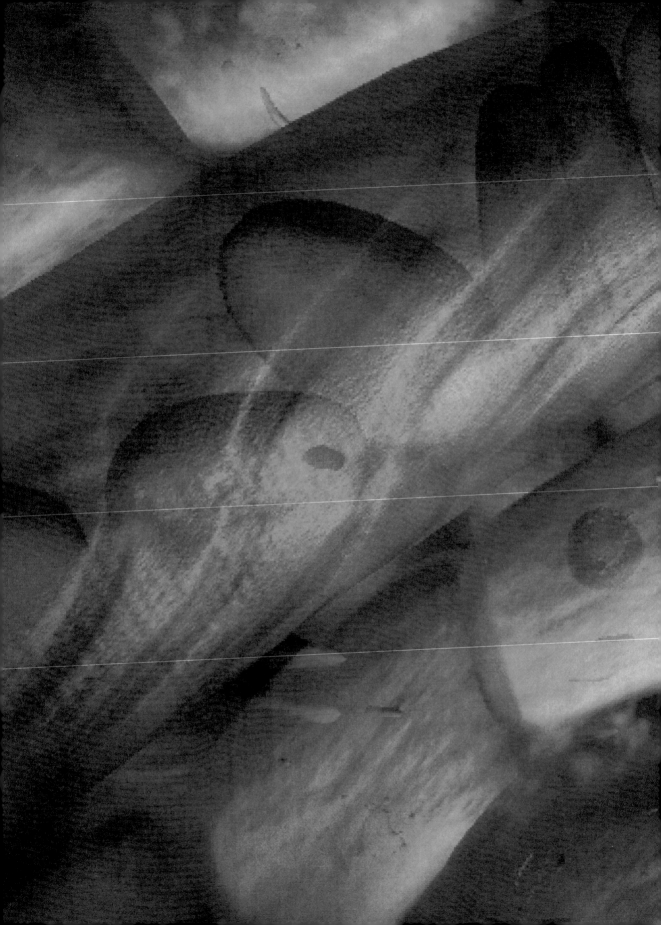

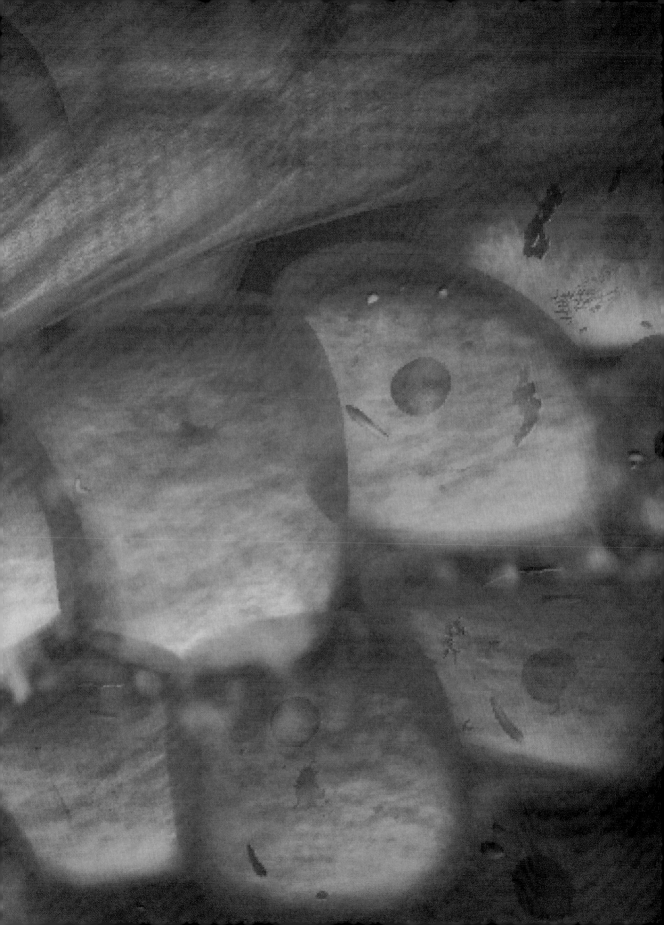

the body's response to repeated injury of the blood vessel wall that leads to atherosclerosis. As teenagers, we all have some plaques in our arteries, but not until our fourth or fifth decade do these plaques start to cause disease. They are the by-product of modern living.

The heart pumps many nutrients to the cells, along with oxygen. Once food is broken down in the stomach and intestines, the breakdown products – fats, sugars and amino acids – are transported to wherever they are needed. But blood is not discriminating – it carries the bad with the good. Cholesterol circulates in the blood at high levels in people who have a high-fat diet. However, despite its poor image, cholesterol is actually of vital importance to the human body. Without it, our cells would not be able to build their protective membranes or essential hormones, such as steroids. The problem is that the body manufactures most of the cholesterol it needs itself – in the liver. A diet high in cholesterol is surplus to the body's requirements.

Cardiovascular disease – disease of the heart and blood vessels throughout the body – kills about 40 per cent of people in the western world. Ischaemic heart disease is the most common cause of death in most industrialised nations. In 1990, coronary heart disease claimed the lives of 489,171 Americans. Of these deaths, 60 per cent were within one hour of the onset of symptoms. Prevention is clearly preferable to cure.

# WHO IS AT RISK?

Certainly, given the statistics above, everyone in the western world is at risk of developing heart disease. A stereotype would be a middle-aged man, overweight, who loves fatty foods and smokes at least a pack of cigarettes a day, leads a sedentary life, has high blood pressure and a strong family history of heart disease. Other, more surprising, risk factors include below average weight at the age of one year, or suffering from a certain type of chest infection. Clearly there is no single cause of ischaemic heart disease, but each of the above risk factors is independently and cumulatively damaging.

### The stronger sex

Our stereotype at-risk man starts with an immediate disadvantage in one respect: he's not a woman. Half the population are born with an in-built protection against heart disease. Simply by being female, they are less likely to suffer from heart disease than men. It has been suggested that the female hormone oestrogen protects blood vessels from narrowing. When the ovaries shut down, at the time of the menopause, the level of oestrogen falls and the number of women having heart attacks suddenly increases. This is true but, just around the same time, the number of men having heart attacks also suddenly increases. Of course men do not have a menopause – at least not one driven by a lack of hormones. But the risk of death from coronary disease accelerates more quickly in men after the age of 50 than in women. By the age of 80–84, the difference in death rate from heart attacks between men and women is eight times greater than it was around the time of the menopause, age 50–54. Therefore women after the menopause continue to be less prone to ischaemic heart disease than men.

However, if post-menopausal women bring their bodies back to a similar hormonal environment as before the menopause, by taking hormone replacement therapy (HRT), they can reduce their risk of cardiovascular disease. This observation strongly supports a role for oestrogen (contained within HRT), as a protective agent against cardiovascular disease. So, should all post-menopausal women take HRT? The answer is still not clear. Although women on HRT are protected from dying of coronary heart disease and have less thinning of the bones (osteoporosis), there may be a price to pay. Studies that have followed up the long-term survival of thousands of women taking HRT have shown that the longer HRT is taken, the greater the risk of developing breast cancer. At present, it seems that the prescription of HRT should be tailored to the individual woman. For example, women who are at high risk of coronary heart disease (e.g. those with high blood pressure, obesity etc.) have a greater benefit from HRT than those at low risk. In other women, this benefit from HRT may not compensate for the fear they have of developing breast cancer and living with its repercussions.

A woman's cumulative absolute risk of death from ages 50–94 has been estimated to be 31 per cent from coronary heart disease and 2.8 per cent from breast cancer. Furthermore, a recent large study showed an overall 20 per cent reduction in mortality in women who took long-term HRT. Although the results of many large trials are awaited, these last two facts suggest HRT is beneficial for most post-menopausal women.

## A surprising risk: low birthweight

An unexpected risk factor for the development of coronary heart disease, not yet proven beyond doubt, is size at birth. Babies who are small at birth and during infancy have been shown to be at increased risk of developing heart disease and high blood pressure (hypertension) in later life. We have long been aware of inherited genetic defects that make us vulnerable to disease in adult life. However, it is a novel discovery that poor nutrition in early life could seal our fate 50 years later, however careful we are with our diet in between.

The evidence for this statement comes from analysis of meticulous infant records, kept by certain health authorities in the UK during the early part of the twentieth century. Analysis of more than 8,000 of these medical records found that an inappropriately low weight at birth or at the age of one year strongly predicts subsequent death rates from coronary heart disease.

## Obesity

Our hapless male stereotype is inevitably overweight. Obesity is now endemic among western countries and makes a major contribution to cardiovascular disease. It is estimated that around 33 per cent of adults in the USA and about 15 per cent of people in the UK are obese. These figures have doubled in the last decade and continue to rise progressively. Despite $30 to $50 billion spent annually on weight-reduction programmes and special foods, there remains the miserable statistic that 90–95 per cent of people who lose weight subsequently put it back on again. Approximately two-thirds will regain their weight within one year and the rest within five years. At any given time of the year, more than one in three of all Americans may be trying to lose weight. Failure brings feelings of guilt and self-hatred. Ever more desperate means are tried, including drugs that may have harmful side effects. For example, a product called fenfluramine with phentermine (fen-phen) was proved to cause serious valvular heart disease; in the year before it was withdrawn from the market, 18 million people had taken it. In any case, whenever the drugs are stopped, weight is promptly put back on.

The body-mass index (BMI) is a useful measure for defining obesity. It is calculated by dividing one's weight in kilogrammes by the square of one's height in metres. A BMI greater than 30 defines obesity, but a BMI of 28 or more is associated with at least a doubling of the risk of ischaemic heart disease compared with the general population. Despite medical encouragement to lose weight and pressure from contemporary fashion to be thin, most people find it almost impossible to achieve, let alone sustain, weight loss.

The reason seems to be related to a system far more complex than a simple conflict between energy intake and energy expenditure. For example, the average man or woman between the ages of 25 and 55 years gains 9.1 kilogrammes in weight. Over 30 years, this represents an excess energy intake of only about 0.3 per cent over energy expenditure. It seems that a person's weight is stable around a set point, which is defended by control mechanisms in the central nervous system. This set point can be overridden by severe caloric restriction combined with rigorous exercise but, when these measures are stopped, the weight returns to the set point. It seems likely that hereditary factors play a role in defining the set point and the slow change it undergoes during a lifetime. There seems little doubt, therefore, that prevention of obesity is far more effective than cure. The way forward is to encourage the young to enjoy regular exercise and eat healthily, without developing fads. Those who are already overweight need help with weight-loss programmes only if they are liable to ill-health, but those who are only mildly overweight need advice on keeping more weight off rather than strenuous plans to lose more.

**Fat of the land**
Of course our at-risk overweight man would happily live on burgers, sausages, anything fried and everything with chips. In countries that survive on a low-fat diet, such as China or Japan, coronary artery disease is uncommon. Further compelling evidence for the importance of diet on death rates from heart disease is demonstrated by the so-called 'Albanian paradox'. Albania is the poorest

country in Europe (gross domestic product – GDP – $380) and has one of the highest infant mortality rates (45 per 1,000 live births). By contrast, adult mortality, particularly death from coronary artery disease, is less than half the rate in the UK (GDP $18,340). Other Mediterranean countries have a similar adult mortality rate to Albania.

It has been suggested that the diet common to these countries contributes to their low death rate from coronary artery disease. This Mediterranean diet is low in total calories, meat and milk products, but high in fruit, vegetables, oils, carbohydrates and wine. By contrast, the average western diet is full of fat, particularly saturated fat – a biochemical description of the type of fat found in butter, cream and cheese. These saturated fats can bind to other fats to form globules that can then seep into the walls of blood vessels.

## The effect of smoking

Our stereotypical man risking his heart must have his regular nicotine fix. The rate of smoking is still generally high and among young women even climbing. It is indisputable that smoking causes lung cancer and accelerates cardiovascular disease. In the USA smoking still accounts for an estimated 200,000 deaths from cardiovascular disease each year. Smoking cigarettes increases the likelihood of a thrombosis within the coronary arteries. Accumulating evidence also suggests that people who live with smokers (or spend a lot of time with them), and therefore inhale their smoke (passive smoking), are also at risk of coronary thrombosis.

## The danger of a sedentary life

If you rest, you rust – so our anti-role model is a classic couch potato. Keeping fit not only helps to burn off excess calories, but increases the level of protective high density lipoprotein (HDL) in the blood (see page 125).

It's never too late to start exercising. There is now good evidence that even elderly people can stave off cardiovascular disease and improve their life expectancy by taking mild exercise, such as walking. The sooner you start, the greater the benefit seems to be.

## High blood pressure

Contrary to popular belief, high blood pressure usually causes no symptoms. Our overweight, smoking stereotype might get occasional headaches, but they are more likely to be caused by excessive television watching! When blood pressure is persistently high for long periods it can damage the retina of the eyes, causing blurred vision, and under these circumstances lead to headaches, but this is the exception rather than the rule. More usually there is a sustained, but more moderate, elevation of blood pressure. This is caused by arteries getting narrower, squeezing down on the blood within; in most cases it is still not clear why this happens. The heart has to work harder to pump blood around the body and becomes even more muscular to cope with the strain. But the blood supply to the extra heart muscle often does not keep up and any existing myocardial ischaemia is made worse.

There is now very good evidence that reducing high blood pressure prevents non-fatal heart attacks and death from coronary artery disease. Furthermore, the drugs for treating high blood pressure are getting better and better. For example, in 1971 high blood pressure was controlled in only 16 per cent of patients in the USA, but now more than 55 per cent are controlled.

## The influence of genes

Our stereotypical at-risk man can't do anything about the genes he has inherited, any more than anybody else can – yet. Perhaps he may comfort himself with the thought that even a lean and fit man may have a family history of heart disease, which makes him succumb to a heart attack at the age of 40. (This thought will of course conveniently ignore the fact that unhealthy habits could bring about that fate irrespective of genes.)

However, fascinating research is now under way that will enable genes with protective properties against ischaemic heart disease to be given to patients who suffer with the disease. These genes could be delivered directly to the narrowed coronary artery, within an artificial pipe; this pipe would be carefully inserted into the narrow part of the artery, supplying the proteins that keep the blood vessel wall open.

## For medicinal purposes only

Some people may be surprised that alcohol has not been listed in the preceding risk factors. After all, it's well known that over-indulgence can cause severe damage. But this is one 'vice' that we can still indulge in, unfettered by any feeling of guilt and soothed by the thought we are doing ourselves good (at least where the heart is concerned).

## Whose life is it anyway?

Why, when we all know that fatty foods, smoking and lack of exercise are likely to harm us, do we not change our habits? Is it the individual's responsibility to change, or should it be imposed from outside? Here's how I see it.

Where smoking is concerned, there's no doubt that peer pressure and irresistible advertising lure the gullible. A *laissez-faire* sense of immortality and a genuine enjoyment of tobacco secure the custom of the rest. Although we have learnt a lot about the genesis of cardiovascular disease, the public have been slow to digest this wisdom. Indeed, it is probably indigestible. I myself find a cigarette at the end of a good meal most pleasurable. Surely just one, now and then, can't do me any harm? I also consume fatty foods. This defiance, while in full charge of my faculties and the knowledge of what harm such habits can wreak, illustrates the difficult task facing public health workers. They have to persuade people less informed than myself to give up something they enjoy. At present we feel fine, we do what we please. In 20 years' time we may still feel fine, but a significant minority of us will be dead or crippled with angina, unable to walk around the block without chest pain or breathlessness.

However, the debate about lifestyle and health risk is in danger of becoming impotent when it infringes upon our freedom of choice. There is no doubt that healthcare workers have a duty to inform the public about the risks of certain habits. But then it is up to the public to make up their minds. Everyone must make a choice. They should not be penalised for making that choice, nor blame anyone else. It is madness to blame cigarette manufacturers for causing ill-health, while their product displays reminders of its harm on every packet. It is out of order for NHS heart surgeons to refuse to perform operations on people who refuse to give up smoking. If patients need an operation, it is often the only way to be free of angina and therefore they must have it. As long as they are fully aware of the risks of continuing to smoke, then it is up to the patients whether or not they choose to stop smoking.

The question of limited resources enters the argument. The waiting list for operations on the NHS is long and full of people in urgent need of surgery. It is felt by some that those who are likely to be more compliant post-operatively are going to give more 'value for money'. Yet what price does the individual who continues to smoke, despite ischaemic heart disease, hold for his life? It is priceless. Life is short and, for some, smoking one of the few pleasures left. The surgeon who refuses to operate applies his own values to that patient.

Those of us with our health still intact also constantly make decisions about risk in life, weighing up benefit against harm: the immediate pleasure of a tasty, high-fat pudding against the distant risk of cardiovascular disease. The latter is too small and too remote to compete with the former. We may feel guilty after such over-indulgence, but here we are getting our wires crossed; it is a guilt born out of vanity, rather than a fear of premature death. As for the young (and perhaps not so young), their youthful arrogance relegates poor health to a level of importance far below that of fashion.

And yet all is not gloom: the healthcare message can get through. In the USA, for instance, although hundreds of thousands of people still die every year from cardiovascular disease, the annual rate has actually been declining since the mid-1960s. During the 1980s, the annual rate of decline was about 3.5 per cent for both men and women. This trend is particularly striking as the rate of death from all diseases (other than cardiovascular disease) did not change during the same period. The declining death rate is due to a combination of falling rates of coronary heart disease and improved survival following a heart attack. Many people have been heeding the message, and giving up smoking and trying to get fit – the two most significant factors for a healthy heart.

Widely prescribed in the nineteenth century for 'medicinal purposes', its medical relevance faded away during the first half of this century. Soon after the Second World War, it was reported that patients dying with cirrhotic livers had an unusually low prevalence of coronary artery disease at autopsy. However, these and more recent reports were ignored due to the belief that alcohol was bad for health. Not until the late 1980s did doctors begin to take the issue seriously. Since then, several large studies have concluded that small to moderate amounts of alcohol reduce mortality from cardiovascular disease by approximately one-third. However, this benefit probably does not come into play until after the age of 45. Younger people who drink alcohol are more likely to die in accidents than those who do not drink alcohol, whereas men over the age of 65 have a progressive reduction in mortality if they drink about three American or four British units of alcohol a day.

The definition of a unit of alcohol varies around the world. In the UK a unit contains 8 grams of alcohol, in Japan 19.75 grams, and in America 12 grams or 14 grams. In the UK, a unit of alcohol is equivalent to a glass of wine, a half a pint of beer or a standard measure of spirits. People who drink in excess of about five units per day increase their chances of dying from alcohol-related diseases, such as cirrhosis of the liver and cancer of the oral cavity, oesophagus and liver, and, as mentioned before, accidents.

The protective action of alcohol seems to work through affecting blood levels of certain proteins associated with cholesterol. Alcohol raises levels of high-density lipoprotein (HDL), which carries cholesterol back to the liver to be used in metabolism or eliminated; so it helps to reduce the level of cholesterol laid down in the lining of blood vessels. Conversely, alcohol lowers levels of low-density lipoprotein (LDL). High levels of LDL in the blood pass through the inner lining of the blood vessel wall. Once inside, it provokes inflammation of the vessel wall and the first stages of a plaque. Cholesterol is trapped in the centre of the plaque, creating a fatty core. It is this plaque that eventually ruptures, causing thrombosis (a blood clot) and a blockage within the vessel. Another benefit of alcohol is that it prevents platelet clotting which we have seen can help lead to atherosclerosis when platelets are attracted to damaged blood vessels. All these actions combine to protect the imbiber from cardiovascular disease in general.

Does the benefit from alcohol apply to all types of alcoholic beverage? The answer is probably yes. However, several studies have suggested wine consumption gives a greater benefit than beer and spirits. However, this is probably due to differences in the pattern of drinking of wine, rather than differences between the actual drinks. In most countries wine tends to be drunk regularly each day, whereas the same amount of beer and spirits are more commonly drunk on one or two days during a week. The action of alcohol on reducing blood-clotting lasts for less than 24 hours and so binge drinkers, unlike regular drinkers, will not have universal protection from the risks of cardiovascular disease.

## ANATOMY OF A HEART ATTACK

Once the lining of the blood vessel is injured, fat molecules can migrate beneath it and cause it to swell. The cap which covers this injured area is a fragile layer of scar tissue. Certain blood vessels are more prone to plaques than others. High-pressure or turbulent blood flow, especially in blood vessels that are continually moving, makes plaque formation particularly likely. Unfortunately, this includes some of the most vital blood vessels in the body, those that supply blood and oxygen to the muscle cells of the heart itself – the coronary arteries.

Coronary arteries can be almost completely blocked off before causing any ill effects to the patient. The body's power to compensate for damage is astounding. If a plaque grows particularly large, almost blocking a main coronary artery, it can stimulate growth of a natural 'bypass' to steer around the blockage.

When a coronary artery narrows, the heart muscle becomes starved of its usual supply of blood (below). This 'ischaemic' heart muscle not only causes pain on exertion (angina) but also stimulates the growth of new coronary arteries (below right).

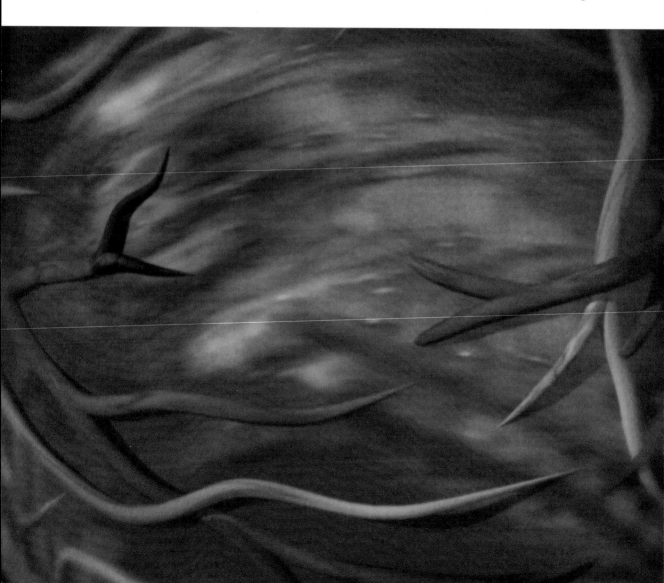

The problem is that these new arteries take at least 48 hours to grow. They are brought on by slowly swelling plaques and the subsequent slow oxygen starvation of the muscle around them. If an artery were to block off suddenly, a self-bypass would not have time to grow.

Arteries to the legs can also become narrowed with atheroma. When leg muscles are exercised, they require more blood and oxygen than at rest. This extra demand for blood cannot be met by the narrowed arteries. The starved leg muscles start to scream out for more blood or for the exercise to stop. This pain that comes on during exercise and is relieved by rest is known as 'intermittent claudication'. If the exercise continues, the body must respond rapidly to compensate. Acting on signals from the labouring muscles, the brain can send out urgent messages direct to the pacemaker cells of the heart. The pacemaker cells respond instantly,

A nerve carries a signal of pain (angina) from a heart muscle with a reduced supply of oxygenated blood.

stepping up the rate at which they fire electrical impulses. The heart rate can double within seconds of this increased requirement. Unfit people have faster resting and exercising heart rates than those with well-trained muscles and atheroma. It is not only the heart that has to work harder. Breathing becomes quicker and deeper to supply the muscles with extra oxygen. The volume of air entering the lungs can increase more than 18-fold.

Hard, sudden exercise in unfit people can do more harm than good. As the heart pumps faster, it forces blood through fragile coronary arteries with increasing speed and pressure. Flowing at up to five times its normal speed, blood will eddy and swirl around uneven artery walls. The thin cap of any existing plaque is put under unbearable strain. It is not necessarily the biggest plaque that will rupture, but the one with the weakest or most inflamed covering. Patients who have had their coronary arteries examined directly and have been found to have minimal narrowings, have gone on to suffer a major heart attack the same day. An apparently trivial plaque may have ruptured, causing a clot that rapidly blocked the whole artery.

When the plaque bursts, the internal response is swift. Sensing injury, blood starts to clot around the rupture. But this normally protective measure, designed to stop the body from bleeding to death, is the last thing the narrowed artery needs. The body's own desperate attempts at repair could now seal its fate. At this moment the patient has no idea of the havoc going on beneath his skin. As the clot grows, less and less blood will be able to get to the area of heart muscle which the damaged artery usually feeds. Downstream of the clot, the drop in blood flow has dire consequences for millions of heart muscle cells. They normally rely on the artery to deliver their food and energy. During exercise, the oxygen supply to these heart muscle cells falters, just when they need it most.

Like the starved leg muscles, heart cells desperate for more oxygenated blood send out pain signals to the brain. This pain is known as angina. During early life as an embryo, the heart develops in the neck before descending down into the chest with its nerve supply. The neck, jaw and arms come from the same region of the embryo and share a similar nerve root as the heart. This is the reason why pain from the heart is often felt as pain in the left arm, jaw or neck. People with angina often mistake the pain for indigestion, while it is relief to someone fearing angina when they discover their pain is merely indigestion.

The body reacts fast to this agonising pain. The brain sends out a volley of signals, triggering a surge of adrenaline into the bloodstream. This is one of the most primitive and powerful reactions in the human body. It happens whenever we are frightened or in pain. Our bodies go into overdrive, to aid our escape.

Arriving at the pacemaker cells, adrenaline instantly boosts the heart. It's as if the body were doing strenuous exercise again. The heart races at over 140 beats per minute. Its task now is to supply oxygen to its own starved muscle. Adrenaline, along with nerve impulses from the brain, prioritises where in the body the blood is needed most. Vessels feeding the skin constrict, diverting blood towards vital organs such as the heart, brain and lungs. The bloodless skin becomes cold, pale and sweaty, as if the patient were running away from danger.

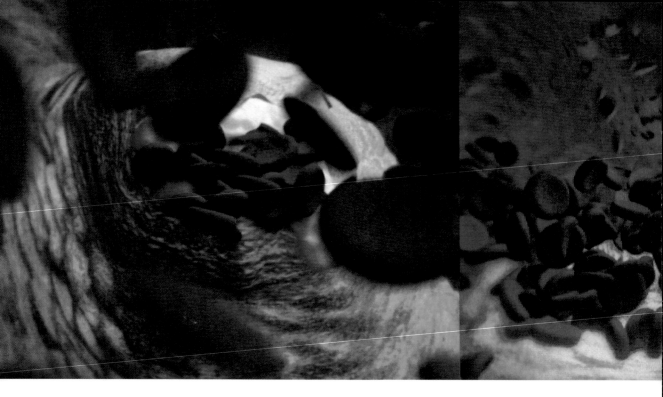

Even now, the heart muscle cells have ways of coping with the crisis. Starved of oxygen, they stop beating and put all their remaining energy into an even more vital task: maintaining their fluid balance. Without this, they cannot stay alive.

Within minutes of the plaque rupturing, the blood clot has completely blocked the coronary artery. This is a 'heart attack'. The heart muscle cells beyond the blockage have now completely lost their oxygen supply. If nothing more is done the muscle cells downstream of the blockage will die – hence the other term for a heart attack, a 'myocardial infarction'. Without oxygen the tiny pumps in the walls of the muscle cells, which normally maintain the fluid balance, start to shut down. The cells begin to swell and eventually become too weak to hold in their own liquid. They burst.

The contents of a ruptured heart muscle cell are released into the bloodstream. Analysis of a blood sample can detect these contents. This is therefore one way to make the diagnosis of a heart attack in a patient with chest pain. An enzyme known as creatine kinase MB and a protein, troponin T, both circulate in high levels after a heart attack. The bigger the heart attack, the more heart muscle cells destroyed, the higher the level of these molecules in the bloodstream and the worse the patient's chance of survival. Blood tests for creatine kinase MB are most widely available, but the levels do not start to show a rise for at least 12 hours after the start of chest pain. This test is therefore reserved for the

## The electrocardiogram (ECG)

Just over 100 years ago, Willem Einthoven (1860–1927) was the first to study the electrical activity of the heart using a string galvanometer. His original ECG machine occupied two rooms and took five people to operate. Some years later, James Herrick (1861–1954) described the ECG appearances of a heart attack. Using more sophisticated ECG recordings, with electrodes on the chest wall as well as on the limbs, it was possible to localise the area of the heart that was damaged following a coronary thrombosis. Nowadays an ECG print-out is generated by a unit no bigger than a textbook and has become integral to the diagnosis of a heart attack.

There are two basic types of ECG machine. One monitors the heart rate and rhythm. This uses three electrodes placed on the front of the chest and gives basic information, such as how quickly the heart is beating and whether it is beating in a normal way. This is the kind of monitor used by paramedics or doctors when they are transferring a patient who is at risk of a sudden change in heart rhythm. Such patients include those suspected of having had a heart attack or anyone who is acutely unwell. This monitor does not tell you whether or not the patient is suffering from a heart attack.

The diagnosis of a heart attack is made by a 12-lead ECG. The patient must lie still, with electrodes on each limb and across the front of the chest. Electrical information from the nerve fibres of the heart can then be recognised by these electrodes which effectively surround the heart. Electrical windows created by dead heart muscle, following a heart attack, can then be identified from the ECG tracing.

The making of a heart attack: when a coronary artery plaque bursts (left) the body responds in the same way as it would if healing a cut on the skin, it forms a clot (centre). Red blood cells and platelets (clotting cells) cover over the plaque and completely occlude the coronary artery (right). Blood can no longer get to the heart muscle beyond the blockage. This is a heart attack. Unless the clot is dissolved quickly, irreversible damage to the heart muscle, if not death of the patient, will follow.

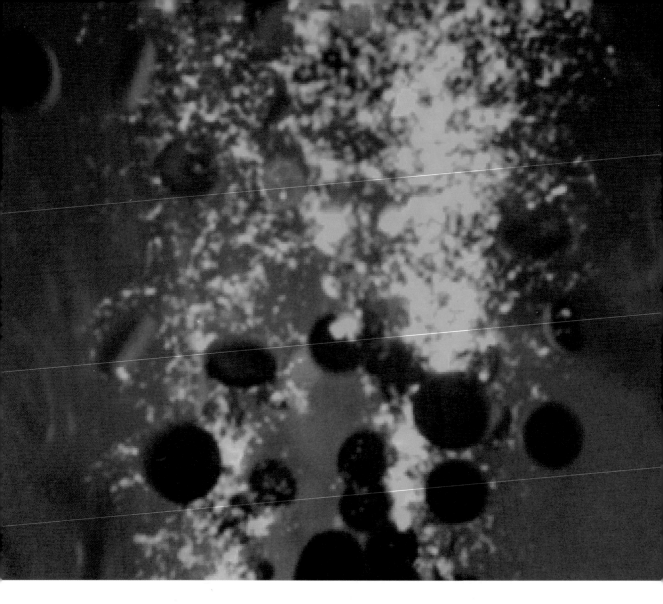

confirmation of a heart attack, rather than the early diagnosis.

The diagnosis of a heart attack can be made quickly by an electrocardiogram or ECG. Until the diagnostic 12-lead ECG (see box, page 131) has been done, the doctors can't be sure whether a patient's coronary artery is blocked. Once they are sure, they must act quickly, otherwise the heart muscle will be irreversibly damaged. Once heart cells have died, they will never be replaced. Time is short.

Before long, the area of heart damage interferes with the function of the whole heart. The injured cells have weakened the power of its contractions. It becomes less and less efficient at pumping blood around the body. The blood pressure drops and as a consequence less blood nourishes the heart muscle. This in turn makes the heart beat even less strongly, dropping the blood pressure even further. To compensate, the heart beats even faster, exhausting the remaining muscle cells even more. The heart is trapped in a vicious cycle. Eventually, a backlog of blood starts to build up in the veins delivering blood from the lungs. This back pressure forces liquid out of the pulmonary vessels, through the delicate lining of the lungs and into air cavities within the lung itself. This is known as pulmonary oedema. If the process doesn't stop, the patient could drown in his own body fluid. One temporary solution to this condition is to get rid of the extra fluid by giving the patient a diuretic (a drug that increases the flow of urine).

As less blood is pumped out of the heart, the blood pressure falls. The brain starts to lose its oxygen supply and mental faculties begin to suffer. The patient becomes confused and disorientated. Although the symptoms of chest pain, breathlessness and confusion all strongly suggest a heart attack, the diagnosis has yet to be confirmed. In the meantime, paramedics can usefully give oxygen, to help the brain function, and pain relief to keep him comfortable. If a heart attack is strongly suspected, then a low dose of aspirin should be given. This acts by preventing the sticky components of blood – platelets – from adhering to the clot over the ruptured plaque.

As the cause of the problem was the body's over-zealous clotting of the ruptured plaque, a drug that can dissolve blood clots is required. Purification of an extract from a bacterium, streptococcus, has produced streptokinase, a powerful clot-buster. Streptokinase dissolves blood clots throughout the body and therefore predisposes the patient to bleeding, especially cerebral haemorrhage. For this reason, streptokinase is not given to patients who have recently had major surgery or are known to have conditions such as a recent bleeding stomach ulcer. Other clot-busting drugs have also now been developed. The sooner the clot-buster is given, then the more effective it is at freeing up the blocked coronary artery. This will reduce the size of the myocardial infarct. It has been shown that streptokinase is beneficial if it is given within 12 hours of the onset of chest pain. However, if it is given within 90 minutes, the patient has a much better chance of surviving. If treatment is left too late, once the heart muscle cells have died off, then the clot-busting drug will be ineffective.

After the death of a region of heart muscle, the heart cannot pump as strongly as before. This leaves the patient less able to exercise; the heart can no longer pump hard enough to deliver blood to exercising muscles. Blood dams back up into the lungs, causing breathlessness on exertion. Fluid on the lung is also suggested by the peculiar symptom of breathlessness while lying flat, but not sitting up – orthopnea. The reason for this is gravity. When the patient lies flat, fluid runs into all areas of the lung, leaving very little dry lung to breathe. But, on sitting up, fluid falls to the bases of the lungs, allowing the upper parts of the lungs to do the breathing.

If the clot-buster is successful, the starved muscle cells will be reconnected with their blood supply. Suddenly they receive a welcome abundance of oxygen. Those cells which have clung to life start to beat again. This is good news. But the early moments of this flood of oxygenated blood are fraught with danger. A heart cell, rescued from the brink of destruction, may recover less well than others. If it

Injection of a drug that dissolves clots such as streptokinase or tissue plasminogen activator (TPA) must be given after the onset of chest pain. Here we can see the clot-busting drug arriving at the coronary thrombosis where it works quickly to pull apart the clogged cells. Arterial blood can then be restored to the starving heart muscle cells.

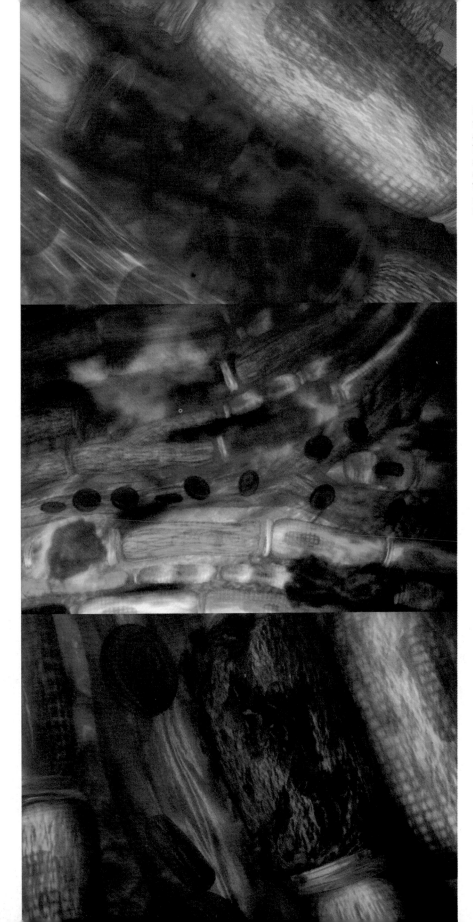

Downstream of a coronary artery thrombosis there is very little, if any, movement of blood cells. The heart muscle cells become starved of oxygen, swell up and eventually die (top and centre). When these cells die they release their contents into the blood (bottom).

beats out of step with other cells, it will become a new source of electrical impulses. The pacemaker loses control and the heart is thrown into chaos. Electrical anarchy breaks out. Every muscle cell in each ventricle starts beating independently. The lack of any electrical or mechanical co-ordination means that the heart stops pumping blood altogether. It is in 'VF' or ventricular fibrillation.

At this time, no blood is leaving the heart and none of the body's organs is receiving oxygen. The first casualty is the brain. If the heart is not made to beat again within two minutes, the brain will be permanently damaged. Until the heart can be forced back into an organised rhythm, cardio-pulmonary resuscitation (CPR) must be started immediately.

If a heart monitor reveals the patient is in VF, it is worth giving a short firm thump to the breast-bone before using the defibrillator. The mechanical energy supplied by this thump is transferred into electrical energy that overwhelms the unco-ordinated electrical activity of the heart muscle cells. This allows a split second during which time the heart's normal conducting system can cut in. The defibrillator machine is a much more effective way of delivering a more powerful electrical shock across the heart. Such a shock is equivalent to the electricity required to light an office block, delivered over a tiny fraction of a second. Similar to the chest thump, the electric shock forces the heart cells to stop beating at once and gives the pacemaker cells a chance to take back control. When VF occurs after the successful use of a clot-busting drug, the patient is usually in hospital and defibrillation can be carried out quickly and successfully. Patients who have collapsed in the street with VF do less well, as they have been in VF for longer and their bodies are already suffering from the ravages of oxygen deprivation.

Having survived the life-threatening VF, the heart begins to repair itself. 40 minutes after the plaque in the coronary artery first burst, it begins to grow a new protective cap. For the next 24 to 48 hours a drug will be given to prevent a blood clot forming at this site again. A low dose of aspirin will also be given every day for the rest of his life. This keeps blood thin, by preventing platelets (the adhesive components of blood) from sticking together. Low-dose aspirin is effective at preventing people suffering a second heart attack, but it is still unclear if it will prevent otherwise fit people from having a first heart attack.

Other useful medications started at the time of an acute heart attack have been shown to reduce future death from coronary heart disease. So-called 'beta-blockers' prevent abnormal heart rhythms and lower blood pressure. If possible,

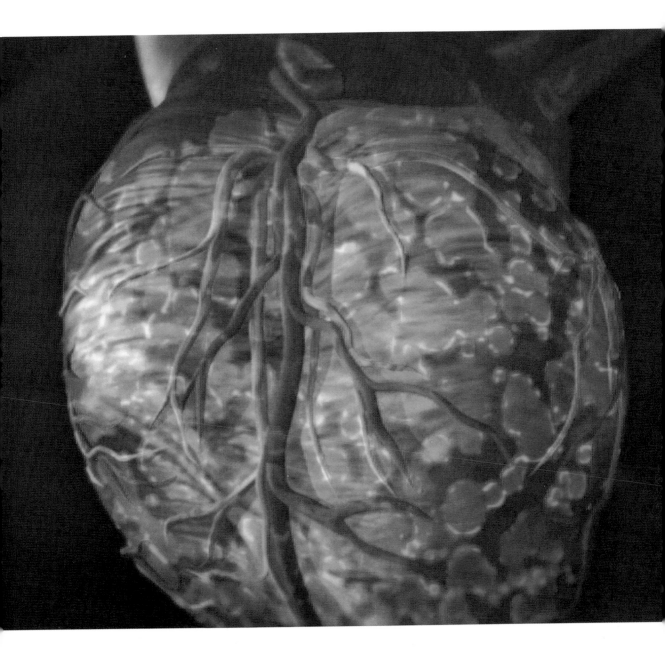

they should be taken for two to three years after a heart attack. Another group of drug called angiotensin converting enzyme inhibitors ('ACE-inhibitors') prevent the heart from failing after a heart attack. They should be given for at least four to six weeks after the attack.

Ventricular fibrillation causes both the right and left ventricles to shiver and tremble without pumping any blood around the body. The patient loses consciousness until a normal heart rhythm returns.

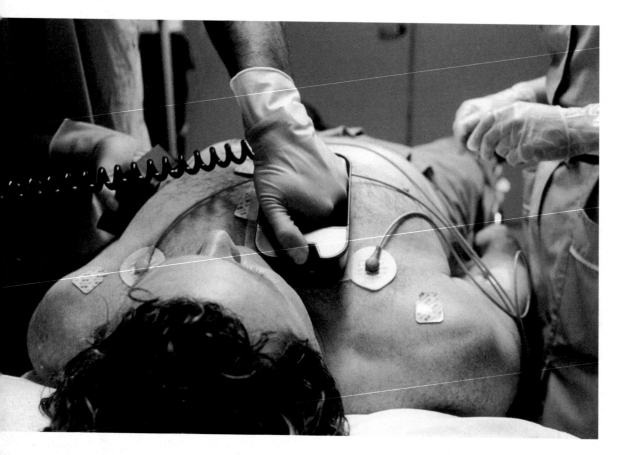

Applying an electric shock to the chest wall – defibrillation – is an effective way of restarting the heart's normal nerve conduction pathway.

## PROGNOSIS

What does the future hold for our patient who has had a coronary thrombosis reversed by a clot-busting drug? If he has a plaque in one coronary artery, he is very likely to have another. Once he has recovered from the initial event, he will have an ECG exercise test. During this investigation the patient is encouraged to exercise on a treadmill or bicycle while having his heart monitored. Heart muscle that shows signs of oxygen starvation – ischaemia – reveals characteristic ECG changes. Such a patient may benefit from a coronary artery bypass graft (CABG).

In order to pinpoint the precise location of the narrowed coronary artery, the patient must have a coronary angiogram. This involves a fine tube or catheter inserted into an artery at the top of the leg. This catheter is pushed up against the flow of arterial blood into the aorta, towards the heart. The origins of the two coronary arteries sit at the point where the aorta leaves the left ventricle. Radio-

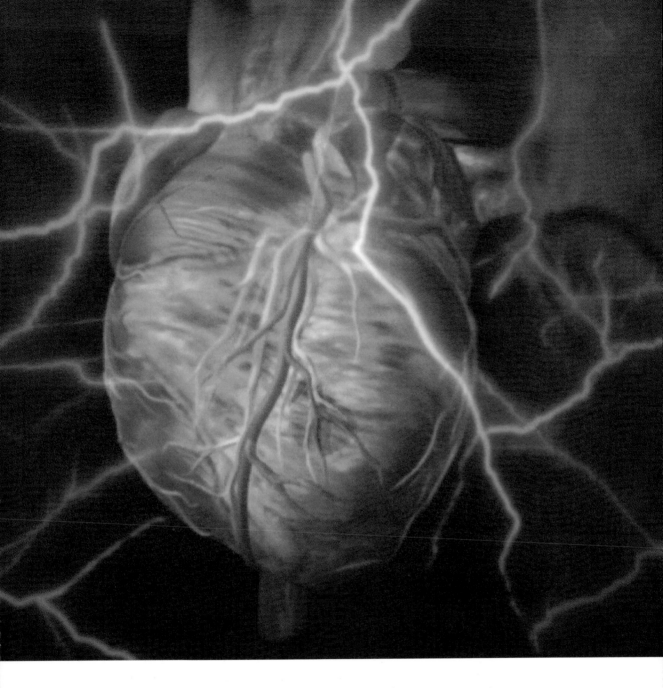

opaque dye can then be injected through the catheter into each coronary artery, in turn. An X-ray film is taken at the moment the dye enters the coronary arteries, allowing any narrowed or blocked vessels to be identified. The cardiothoracic surgeon will then know which vessel to bypass.

Around the world, almost one million people each year receive some form of procedure to improve blood flow through or around a narrowed coronary artery.

Here the heart is shown having received a powerful electric shock from a defibrillating machine. This interrupts the electrical activity of VF and allows a split second for the heart's normal conduction system to cut in.

## Defibrillation

I have defibrillated hundreds of patients and seen both good and bad outcomes. However, my first time was a disaster. As a new houseman on the cardiology ward I had mastered the theory but not the practice of defibrillation. A 70-year-old man with severe ischaemic heart disease went into VF at 4.30 in the morning. I ran to his bedside where the defibrillator was being plugged in by the nursing sister. Others were doing CPR. I agreed with the diagnosis, charged up the machine to 200 joules and placed the paddles on his chest. Everyone stood back and I fired off the defibrillator. There was a loud explosion and a small fire started on his chest. I had inadvertently placed one of the paddles on an electrode that was monitoring his heart. The fire was quickly patted out and a further shock of 360 joules, with more carefully placed paddles, fortunately did the job. On that occasion it wasn't only the patient who got a shock!

Not surprisingly, even when it's done properly the first time, most people are a little dazed after defibrillation. They will have been unconscious due to a reduced oxygen supply to the brain and distracted by the discomfort of CPR. Some patients have recurrent VF, which needs to be suppressed by drugs. Others are resistant to such treatment and in time may require a small, implantable defibrillator. These devices are still being refined, but for certain patients they are of undoubted benefit. They are larger than an ordinary pacemaker and use up their batteries much more quickly. Furthermore, some of the early implantable defibrillators had a habit of firing off inappropriately, giving the patient quite a fright!

During a coronary artery bypass operation, there are effectively two 'bypass' procedures taking place. The first bypass allows the surgeon to operate on the heart. By inserting a tube into the right atrium, venous blood returning from body is pumped through an oxygenating machine and returned to the aorta, bypassing the heart and lungs. The heart is then stopped by direct injection of a very cold fluid – cardioplegia. The surgeon is then free to get on with the coronary artery bypass graft. A length of vein is removed from the leg. One end of the vein is stitched into the base of the aorta; the vein is then cut to length, before being stitched in on the far side of the block. Blood can then flow freely to the ischaemic heart muscle via the vein, bypassing the blockage. This may need to be done on several branches of the coronary arteries, hence 'triple' or 'quadruple' coronary artery bypass grafting. The heart is then restarted with an electrical shock and the tubing to and from the oxygenating machine removed. It is interesting that the thin-walled vein in the leg can take the high pressure, arterial blood from the aorta. Indeed, these veins slowly adapt to their new environment by developing a thick wall, similar to an artery.

CABG is clearly a big and risky operation. After much refinement, only about one in 100 patients do not survive it. This is usually because the heart cannot be restarted at the end of the operation and the patient dies on the operating table. Following a CABG, about five per cent of patients have some sort of neurological damage, ranging from a stroke to depression. It is an operation that is therefore reserved for the most severe cases of coronary artery disease. If only one coronary artery is narrowed, then the obstruction might be relieved by a far less invasive technique, 'balloon angioplasty'. A special catheter with a built-in balloon is sent up the aorta and into the coronary artery with the balloon deflated. When the catheter reaches the narrowing, the balloon is inflated and the narrowing flattened against the wall of the coronary artery. Occasionally the balloon is blown up too much and the coronary artery ruptures. This is a catastrophe. Emergency coronary artery bypass surgery must be performed to save the patient's life.

## Cardio-pulmonary resuscitation (CPR)

Successful CPR should be carried out in a methodical manner, step by step. Once the patient has been found to be unresponsive, help must be called for immediately. The next vital steps can be remembered as 'ABC'. A is for airway, which can be obstructed by a foreign body. A finger must check that the back of the mouth is clear and the tongue forward. The airway can then be opened by extending the neck and bringing the jaw forward. B is for breathing, using air expired from the mouth of the resuscitator directly into the mouth of the victim, or indirectly via a mask. The nose must be pinched to prevent escape of air. C is for circulation, generated by compressions of the chest 60 times a minute. Lung inflations must be co-ordinated with chest compressions, at a ratio of 1:5 if there are two rescuers or 2:15 if the rescuer is single-handed. This process should continue until ventilation and circulation are restored, technical support arrives, or the rescuer is too exhausted to continue.

The long-term outcome for patients with a normal body temperature, who have had CPR for longer than 20 minutes, is very poor. There are, however, amazing stories of people, mainly children, who have been successfully resuscitated more than one hour after a cardiac arrest in very cold conditions. The low body temperature allows the function of the brain to slow right down, like a hibernating animal, and therefore its demand for oxygen falls. The low oxygen concentration in the blood of someone whose heart has stopped is therefore tolerated for longer by the cold (hypothermic) patient. An important rule, when considering stopping resuscitation, is 'never pronounce someone dead, until they are warm and dead'.

Most of us have seen images of rescuers performing CPR. They look good, but are they doing any good? They heroically press down on the chest wall, often breaking ribs with over-exuberance. They fill the lungs of the victim with foul air, rich in carbon dioxide. What exactly are they doing? By pushing the front of the chest against the back, a pressure gradient is created between blood vessels inside and outside the chest cavity. During a compression, blood flows into the aorta and, on relaxation, blood flows into the pulmonary circulation. The ventricles are not compressed at all.

During these heroic efforts, rescuers generate a blood pressure less than 25 per cent of normal. If the heart is to have any chance of restarting, then the blood flow to the heart muscle itself needs to be excellent. Yet chest compressions generate only five per cent of the expected blood flow to the coronary arteries. In the vast majority of cases we are probably not doing much good. But in a few, the effort might buy enough time for the medics or paramedics with equipment and drugs to take over more effectively.

The ultimate heart operation, reserved for those with a heart that has lost its ability to pump, is a heart transplant. First performed in 1967 by the South African surgeon, Christian Barnard, this operation is now performed daily, in multiple centres around the world. Drugs used to suppress a recipient's immune system are becoming increasingly effective, making rejection less and less likely. A further problem is that patients with a failing heart may not be able to survive until an appropriate heart becomes available. There are artificial pumps that can be used to assist the failing heart, but this is only a temporary salvation. Much effort is currently being placed on the development of 'artificial hearts'.

The lack of organ donors has pushed along research into the use of animal organs in humans (xenotransplantation – transplantation across species). We are already able to use heart valves from pigs, as they are fairly inert structures. However, the pig has a very different immune system from the human. So far, attempts to use bigger organs, such as the pig heart or kidney, have resulted in

Once the immediate emergency
of a heart attack is over, the
patient and his doctors need to
assess his risk of suffering
another one.

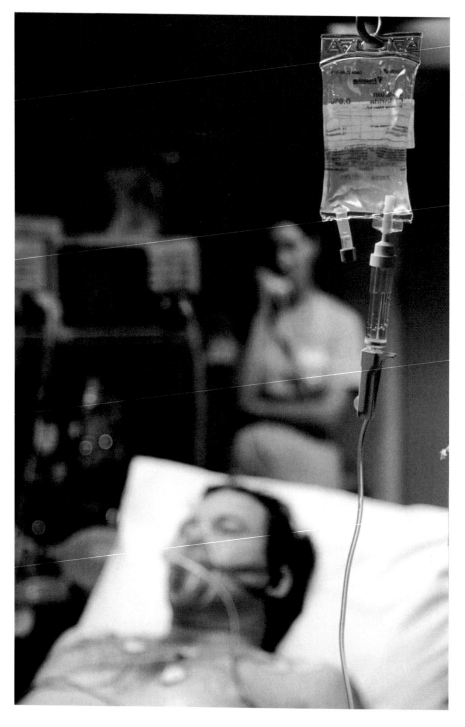

destruction of the organ within minutes. Amazingly, this was first attempted in 1902, by Emmerich Ullman. He attached a pig kidney into the blood vessels of a young woman dying of kidney failure. Not surprisingly, the kidney was rejected within minutes.

There is some evidence that certain infections can cross species, causing new diseases in the recipients of organs from other species. It is a possibility that must be fully explored before xenotransplantation is truly viable. Meanwhile, important progress is currently being made to overcome the immunological barriers of xenotransplantation. Research may lead to this becoming a likely lifesaving operation for patients with heart failure within the next 10 years. The social and ethical objections to this form of treatment are also nearly overcome. Put in the context of mass animal culling for food, the few animals culled for use of their organs to save human life seems a very small price to pay.

Whether a new heart would be human or animal, or purely mechanical, or for now the old one has been saved and beats on, however successful the treatment has been, though, some heart cells are likely to be lost for ever. Those that survive but are injured suffer a sort of 'concussion'. Within a month they are usually back to beating as normal. Until that time, the healthy heart muscle can take on the extra work. However, tests may reveal other areas of the heart at risk from coronary artery disease. From now on the job of supplying the body's 100 trillion cells with blood will be harder: the heart is more than ever under pressure.

# 6

# SHUT DOWN

The only certainty in life is death. However not all of us are destined to grow old. We fulfil our biological purpose by maturing, reproducing and caring for our young. Once this is done, we are dispensable. Yet more and more of us are now surviving until old age: it has been estimated that of all humans ever to have survived until the age of 60, more than a quarter of them are alive today. Why do we have to age? Can we live for ever?

**Previous page:** The structure of our DNA is broken by damaging molecules known as free radicals. We have an inbuilt system to combat the action of free radicals, cells often commit suicide once irreparably damaged. However, over many years of free radical attack the number of cells in our individual organs diminishes and consequently, we age.

So far, this book has dealt with the beginnings of human life, its growing up, the complexities of our bodies and minds and the accidents and illnesses that may befall them. We have looked at our unconscious drives and our rational thoughts, and overall at the marvellously intricate systems that have evolved to sustain us. This final chapter deals with the last stages of life and the process that ends it.

Of course, we are liable to die at any time throughout our lives – from the very moment of conception we begin a long struggle to stay alive. In the womb more than half of all early pregnancies are lost but, if we survive until birth, the vulnerable first few days of life still claim a high number of young lives. Over the next few years, our clumsy bodies and our inexperienced minds put us into dangerous situations, and not all of us make it through. At any time, our immune system can be caught off guard and overwhelmed by an invading organism. Just 100 years ago, those who survived into adult life were the lucky ones. Even today, in many parts of the world this is still the case.

To understand how and why our bodies age, we will follow an elderly man on the last day of his life. He is 87 years old: a ripe old age at any time and in any country. He fulfilled his biological purpose – to reproduce – long ago; in fact, his son now has a young son of his own. As far as nature is concerned, this grandfather is living on borrowed time, and has been doing so for decades.

## LIFE EXPECTANCY

In terms of living longer, there has never been a better time to be born than the present. Despite all the damage we inflict on our environment and each other, our chances of living into old age are improving all the time. But this statement holds true for only some of us: those who live in the industrialised west rather than the 'developing' world. The reasons for this disparity are the same as those given to explain the increase in average life span seen in industrialised nations over the last 100 years. That is, improved levels of sanitation and nutrition rather than any specific medical advances. For example, the advent of antibiotics has had a small impact on our improved survival, even against infectious diseases, in comparison to better housing, an efficient sewage system and a reliable supply of wholesome food. Added to this, education programmes and a reasonably settled social structure improve the chances of survival of any population group. Therefore, in many parts of the world with few or any of these advantages, the people suffer

More people are living into old age than ever before. By 2030, more than one quarter of the British population are likely to be over 65 years old. To live even longer than this depends on the genes you were born with and your lifestyle.

accordingly, their lives inevitably shortened. Natural disasters such as crop failure and extreme climatic conditions are exacerbated by human corruption and, very often, warfare. In famine-struck Ethiopia, for instance, the average life expectancy is barely 41.

Our old man has grown up in the affluent west, in Britain, but until very recently his chances of living as long as he has would have been tiny. Throughout history there have always been people who lived to a great age, but they were very much the exception. At the time of the Roman Empire, for instance, the *average* life expectancy for a European was 20 years. By the year 1000 it had risen to about 30. By the mid-nineteenth century, one could expect to live for 39 years. By 1911 the figure was 46 and in 1930 it was 55. In Britain, over the past 100 years the proportion of the population over 65 has risen from five per cent to 16 per cent.

It is not always understood just why the average life expectancy has increased. Of course, modern medicine can keep old people alive longer than before, but the increase is essentially, and ironically, due to our ability to keep young people alive rather than the elderly. As standards of living have improved, fewer newborn babies are now dying and fewer young people succumb to infections. These young deaths used to skew the average life expectancy to a lower age. For example, of all the children born in Manchester in 1830, only half of them would survive to the age of ten. By 1919, 12 per cent of deaths in Britain were in the first year of life, while now these are just one per cent. (Even so, I recently saw a notice in a neonatal unit that reminded us about the dangers of the first few minutes of life. In bold letters it stated, 'The first five minutes of life are the most dangerous.' Underneath some wit had written, 'The last five minutes are pretty dodgy too!')

So all the improvements in our modern life in the industrialised west – medical as well as environmental – have had more impact on the life of the young, rather than extending longevity itself. An important distinction must be made between the average life span, which has increased dramatically over the last 100 years, and the greatest age you are likely to reach, which has increased very little. For example, in 1890 a man who reached the grand age of 65 years could expect to live another 9.5 years, compared with a 65-year-old living in 1990 who could expect to live another 13 years. The longer you live, the greater your chance of exceeding the average life expectancy. If you are, say, a British woman just celebrating your seventy-seventh birthday, and you read that the average life expectancy for women in your country is 77.4. You might think you've got only

## The longer-lived sex

Throughout life, men are twice as likely to die as women. Although women show the same doubling of mortality rate as men – every eight years (see box, page 150) – fewer women die than men at any given age. This is true even before birth. There are thought to be twice as many males as females at the time of conception, but there are roughly equal numbers of both sexes at birth.

Males are also more likely to kill themselves during adolescence and up to about the age of 23. During this period, the death rate of a maturing male (usually from accidents and suicide) is three times higher than that of women. It's a period described by one academic as 'testosterone dementia'.

Behavioural factors apart, women have greater protection from the most common killer, heart disease, through the influence of the female sex hormone oestrogen, on the heart and blood vessels. Yet males also die from cancer and some infections more frequently than women. This appears to be due to a more powerful immune system in women and may explain why auto-immune diseases (diseases when the immune system is turned against its own tissues) are more common in women.

The overall result is that women live longer than men in almost every country in the world. In the UK the average man can expect to live for 71.5 years, whereas the average woman may live for 77.4 years. It is no surprise that the oldest person ever to live (and with reliable records to prove it) was a woman: Jeanne Louise Calment was born in 1875 and died in 1997 at the amazing age of 123.

another five months to live – but in fact the statistics predict another 10 years of life for you. The average figure must always take account of all those women who die at various ages before 77.4.

The increased average means that more and more of us are living longer; and as we do so, we are developing the range of physical and mental changes that have always been associated with ageing. The impact of one important change in later years has often been underestimated: retirement. When we retire from regular work, we lose the daily routine that has been with us for decades. This disrupts well-established patterns of behaviour and has a knock-on effect that disrupts our internal biological clocks (see Chapter 3). Daytime naps lead to disrupted sleep patterns at night, and fading light perception from cataracts disrupts established internal rhythms. Melatonin release is less cyclical and sleep begins and ends earlier. The synchronisation of many biological functions within the body is upset, making contact with the environment and its community even more difficult. Of course, many pensioners manage to adjust and thoroughly enjoy their well-earned retirement; but the lives of others are blighted. Only now, with so many people reaching this stage of life, is our society experiencing the full impact of ageing.

# WHAT IS AGEING?

The process of ageing can be described in a theoretical way, using a mathematical formula (see box, below), but this may mean little to people who just want to understand how bodies change physically as we get older. The statisticians may tell us that, according to their formula, we start ageing from the age of 11 – but how can we relate that to real life? A child may be getting older, but surely not 'ageing', with the connotations of physical decline?

From a biological standpoint, it would seem more sensible to suggest ageing begins once we have reached our maximum physical and reproductive capacity – around our mid twenties. But not all parts of our body age at the same rate. It appears that each of our organs has its own ageing agenda. For example, when a female baby is born, her ovaries contain all the eggs she will ever have – and vastly more than she will ever need: millions rather than hundreds. By puberty she will have around 250,000, a number that declines until the menopause, when no more eggs are left. Can it be said that the ovaries of a baby are ageing?

Different organs not only age at different rates; they age in different ways. Some of the major organs have their full complement of cells at birth – the brain, the heart and the kidneys. Ageing in these organs means a gradual loss of cells, leading eventually to impaired function. Other important organs, such as the lungs, the skin and the liver, have the capacity to replace dead cells. Tissues all over the body have their own stem cell 'factories', which keep up a steady replacement of new cells. Normally, 200 million red blood cells are replaced every day; the lining of the stomach is renewed in a three-day cycle; and our bones are completely replaced every ten years. But as we age this process slows down. Slowly, stem cells become defective and die.

## Ageing by numbers

When does ageing begin? One way of defining it is the mathematical approach, where death rates are calculated for different ages. For example, in the developed world our chance of dying is relatively high in the first year of life (one in 1,000); it declines until the age of about 10–12 years, but thereafter increases relentlessly the older we become. So ageing could be said to begin when we are about 11 years old.

A mathematical analysis can also be used to describe the ageing process once it has begun: the time it takes for our likelihood of death to double. In most developed countries this is around eight years. For example, at 38 years of age we are twice as likely to die as we were at 30 years of age, but half as likely to die compared with people aged 46 years, they in turn are half as likely to die than people aged 52 years, and so on. It is interesting to note that both men and women age in this pattern, although women live longer (see box, page 149).

## Radical attack: threats to cell life

Cells are at risk not only from external factors – especially strong sunlight and pollution – but also from their own internal metabolism; they generate their own pollutants. Inside every cell in the body are tiny electrically charged particles, known as free radicals. Free radicals are the waste product of healthy cell activity. Highly unstable, they have the potential to damage anything they collide with.

In the nucleus of a cell is its genetic material, made up of DNA. A cell's DNA suffers around 10,000 free radical strikes a day. Each time the DNA repairs itself, but on rare occasions the repair process fails. When a cell's DNA is damaged, it becomes a liability. It could start behaving in a way that endangers

Stem cells are the factories that produce replacement cells for those lost or damaged. When stem cells die they are not replaced.

life. The body has evolved a drastic defence mechanism to prevent this happening: damaged cells are programmed to commit suicide.

Our skin, the largest organ of our body, provides a particularly graphic demonstration of the way in which cells are killed and then replaced. Skin is usually exposed to ultra-violet light from the sun and, in urban areas, high levels of air pollution; masses of cells are lost every day. Unwise sunbathing vastly increases the damage. Under seven layers of skin are the stem cells that generate masses of new cells in order to keep up with those lost. But, constantly battered and destroyed from outside, even stem cells themselves eventually die. Skin cancer becomes more common the older we get.

Cancer is more common in some organs than others. The reason for this is that cancer develops following a fault in cell replication and therefore occurs more frequently in organs with actively dividing cells. For example, the cells of our brain, heart and kidney are not replaced after birth, and these organs only very rarely develop tumours. Those tumors that do unfortunately occur are often found

## The rate of living

Animals all have different rates of ageing. Mice live about two to three years, cats about 10–15 years, and dogs about 10–20 years. It is often said that one human year is equivalent to about seven dog years. To see a dog, aged 13 years, limping and drooling around a park at an age when humans are breaking into puberty, must make us wonder what it is that accelerates the ageing process in these animals, and what slows it down in us.

In general, big animals live longer than small ones; small animals, with fast metabolic rates, burn out and die more quickly than those with slow metabolic rates. Early workers in this field called this 'the rate of living' theory. The faster our bodies work, the shorter our lives. This work was given credence in the late 1950s by George Sacher, a US biologist and statistician. He found a direct correlation between a mammal's weight and its longevity, and that the faster the metabolic rate, the shorter the life. The rate of living theory was further supported by firm evidence that laboratory rats fed on a restricted diet live longer than those allowed free access to food. The less the rats eat, the lower their metabolic rate, and the longer they live (up to the point where calorie restriction becomes frank starvation). Furthermore, these calorie-restricted animals have fewer tumours and remain more capable mentally and physically than their fatter peers.

The normal products of oxygen metabolism in a cell are increased in those animals with a high metabolic rate. It is these oxygen by-products (free radicals) which have been linked to ageing. Free radicals can destroy large molecules in the cell such as proteins and DNA and stop it functioning normally. To combat this damaging process, cells produce anti-oxidants; but as the body ages, the cumulative effects of oxygen free radicals cause its decline (see box, page 163). It is suggested that the rate of ageing depends on the amount of stress from free radicals that an animal has to endure. For example, hibernating or cold-blooded animals and those who eat fewer calories have lower rates of metabolism and less cell damage from free radicals, and therefore these animals live longer.

But this cannot be the whole story as there are so many exceptions to the rule. Wild bats, for example, are no bigger than mice, yet they live up to 30 years. Even those bats that do not hibernate have a fast metabolic rate and live for decades longer than predicted by the 'rate of living' theory. So if the rate at which we live our lives cannot explain ageing, maybe it is the way our cells combat the free radical stress that is important; that is, long-lived animals have more effective inbuilt anti-oxidant systems (see box, page 163).

in childhood, left over from the time when foetal cells were actively replicating.

It may not just be damage by free radicals that causes cell suicide. It is possible that cells can only divide so many times before they become defective and kill themselves. An early orthodoxy said that there was no reason why cells could not in theory carry on dividing for ever. In fact, experiments on normal human cells taken from a volunteer and grown in a culture medium in a laboratory only divide a certain number of times before dying. This is unlike cancer cells, which can grow for ever. Indeed, descendants of cancer cells taken from a young woman who died in 1951 are still used for research purposes in hundreds of laboratories around the world today.

It used to be thought that ageing consisted precisely of this failure of cells to divide for ever, a failure which was regarded as a pathological degeneration. But in fact it is part of normal cell function that genes are turned on which stop cell division. Such genes are of clear benefit in foetal growth: liver cells, for example, are supposed to grow enough to create an infant's liver and then stop, not to create a newborn baby consisting entirely of liver. At any stage of life, though, uncontrolled cell division (cancer) can be lethal. So this limit imposed on cell division is a normal physiological process. It is not the same as, or the cause of, the ageing process.

## THE EFFECTS OF AGEING

Most people, if asked whether they minded getting older, would agree with the view that they don't mind if the alternative is death! But there's an important difference between getting chronologically older and the process whereby mind and body become progressively worn out. It isn't always easy, though, to separate the effects of pure ageing from the influence of other factors – disease, accidents and environment. The challenge these days is not only to stay alive for longer, but to maintain health to as high a standard as possible.

It is clear to any remotely observant person that people age at different rates. There are several reasons for people being 'well preserved'. First, they are lucky with the set of genes inherited from their parents. Barring accident, they escape the onset of disease in early and mid-life and pass into late life with the ageing process making only a slight impact on the function of their organs. But, just as at any stage of life, both physical and mental deterioration can be slowed down by keeping the body in good condition. Physical and mental exercise, along with a

As we grow older, we become less able to conserve body heat. Even in industrialised countries, old people still die in their homes from hypothermia.

careful diet, can keep a 70-year-old in better shape than a slothful person half the age. It also helps if living conditions are of a reasonable standard, particularly where heating is concerned.

A thermal image showing how our peripheral parts suffer in the cold: our hands, ears and nose.

Every cell, regardless of its function, has one thing in common: its act of living generates heat. This heat helps to sustain life. With a dwindling number of cells, our old man's body produces less heat than it once did. Worse still, his body is losing its capacity to conserve heat. Our body temperature is controlled by a thermostat that has receptors throughout the body that can detect changes in temperature. Messages are sent to the brain which can then adjust blood flow through arteries near the skin. When we are cold, blood vessels near the skin constrict, to prevent heat loss from our warm blood circulating within. But, as we get older, our arteries become less able to constrict. Like old pipes, they have furred

## Clues to ageing: an ageing gene

It has been thought at various times that a particular gene, or its mutation, could hold the key to ageing. One such was a very rare condition caused by a mutation in a single gene: 'progeria'. Babies born with progeria look normal, but from then on they grow very slowly and in a way that makes them look decades older than they really are. They live until their early teenage years and then die suddenly. On the face of it this would appear to be compelling evidence for an ageing gene, but this is not the case.

Unlike a chronologically aged person, children with progeria do not show signs of ageing in all of their internal organs. Indeed, it is only because their skin is thin, their hair scanty and their facial features wizened that they appear old. They die prematurely through atherosclerosis and heart disease (see Chapter 5). The gene causing progeria is therefore responsible for only a few features of ageing; it does not, however, explain the overall picture. Indeed, many other gene mutations also cause individual signs of premature ageing, such as premature greying or pre-senile dementia.

Down's syndrome is a genetic disorder characterised by an extra chromosome (number 21). People with Down's syndrome show some signs of premature ageing, because the extra chromosome contains the gene that produces a protein known as amyloid. Excess amounts of this amyloid protein have been found in the brains of people with Down's syndrome and age-related Alzheimer's disease. However, this feature of ageing in Down's syndrome is not accompanied by the full range of ageing characteristics seen in elderly people. Therefore, the gene for ageing doesn't lie solely on chromosome 21; no single gene individually explains the process of ageing.

Except perhaps a recently discovered gene in a worm, *Caenorhabditis elegans*. A mutant form of this worm, which lacks a gene that normally produces an enzyme to protect it from the toxic by-products of cell metabolism, ages much more rapidly than normal worms. The human equivalent of this enzyme is also faulty in a rare group of diseases that causes precocious ageing. This discovery supports the biological explanation for the 'rate of living theory' of ageing (see box, page 152) and might explain why some animals – which may have lower levels of the protective enzyme – live shorter lives than would be expected for their metabolic rate.

up with excess fat and calcium. Their elastic walls have stiffened and so more heat is lost to the outside world. This makes old people particularly vulnerable to hypothermia – every winter in Britain brings a sad litany of impoverished pensioners who cannot manage to keep warm enough to live.

### Use it or lose it: the ageing brain

We are born with all the brain cells we will ever have – and, as with a baby girl's eggs, a super-abundance. Throughout our lives we lose many thousands of brain cells each day, which sounds alarming until compared to the total number we have at birth: over 100 billion. By the age of 87, the number of brain cells we will have lost amounts to only about 10 per cent of our total starting number. Despite this steady loss of brain cells, we can hardly claim to be at our mental peak at birth, so how can we reconcile brain-cell loss with mental ability? The answer is in our ability to develop new skills, a time when brain cells reach out and make new connections with each other – trillions of them.

We may be getting older, but our intellectual function doesn't necessarily

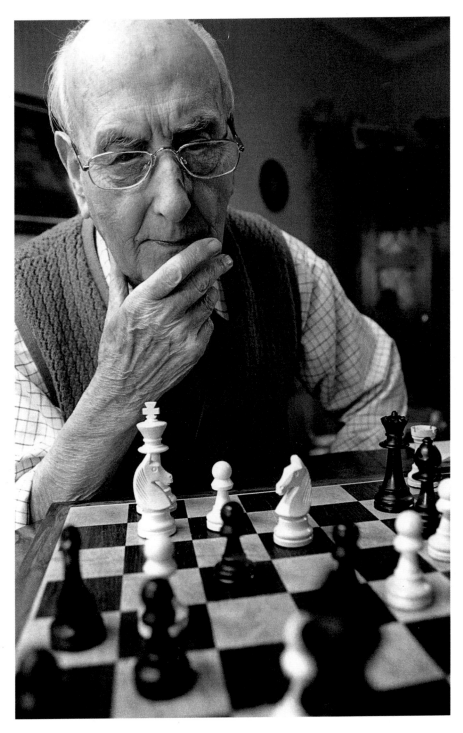

One of the best ways to keep mentally alert as we grow older, is to keep using our brains. Learning new skills leads to more new connections between brain cells and slows down any decline in mental function.

It has been estimated that we lose up to 50,000 brain cells each day. Although we consider all our brain cells to be precious, we probably only lose a small percentage of our total 100 billion brain cells over a normal life time.

decline. Our elderly man is still mentally sharp most of the time. He keeps his brain exercised, playing challenging games of chess, and taking an intelligent interest in new (for him) inventions like the computer. Older people who remain mentally alert and can learn new skills at an advanced age can still lay down new connections between brain cells. We often hear stories about grandmothers who learn computing to keep up with their grandchildren, or grandfathers who catch up on the education they were denied in childhood. The more you use your brain, the more connections it makes and the more capable it becomes.

This is not to say that the brain never flags. The passing years do eventually make it more difficult for the brain to lay down those vital connections.

Certainly the speed at which nerve impulses are conducted slows down by about 15 per cent between the ages of 30 and 80, and consequently reaction times become progressively slower. It has therefore been argued that our brain ages when we start to lose the connections between brain cells. But because we are born with far more brain cells than we ever use or need, the effects of a shrinking brain take a long time to show. Indeed, at 87 our man shows no sign of impaired reasoning – though he does suffer a common complaint of old age: a failing memory. This is because cells do not die evenly in all areas of the brain.

A life's experiences are translated into memories in certain parts of the brain, notably the hippocampus. The cells here die at a faster rate than the rest of the brain; since our old man was 50, he has lost as much as a fifth of the cells in his hippocampus. While it has no trouble following well-worn paths, his brain is becoming less and less efficient at laying down new memories. This may explain one of the apparent paradoxes of old age: why elderly people can't remember recent events as well as they can recall experiences from their distant past. Our old man may exasperate his young grandson by forgetting what he has just been told, yet be able to remember events in his own boyhood with crystal clarity.

## Losing the mind: dementia

Dementia is a progressive loss of our intellect, memory and personality. About five per cent of all people over 65 have some degree of dementia but, as we grow older, the numbers affected increase. Every five to eight years after the age of 65, the number of people with dementia doubles; so that 20 per cent of 80-year-olds are affected. This of course means that 80 per cent of the over-80 population preserve all their mental faculties. Furthermore, the number of people with dementia does not climb exponentially after 80, but flattens out. Occasionally, dementia is mistakenly diagnosed in those who are deaf or have a mild degree of forgetfulness – both common accompaniments as we grow old.

There are several causes of dementia, but by far the most common is Alzheimer's disease. This was first described in 1907 by a German neurologist, Alois Alzheimer (1864–1915). Originally it was a diagnosis reserved for those people with early-onset (pre-senile) dementia, but now it is recognised in all patients whatever their age, who have dementia associated with a type of protein (amyloid). Amyloid steadily accumulates between brain cells, forming a so-called 'senile plaque'. There are unfortunately inherited forms of Alzheimer's disease in which the gene that produces amyloid is abnormal.

A CAT scan of the brain will pick up widespread evidence of brain shrinkage, or atrophy. However, Alzheimer's disease has its greatest impact on the nerve pathways to and from the hippocampus, the area of the brain that has a major role in storing memories. The hippocampus itself remains relatively untouched by Alzheimer's disease, but it becomes isolated from the rest of the brain by progressive destruction of its nerve connections. Therefore one of the most striking symptoms of this condition is severe, progressive memory loss (amnesia).

There is no specific treatment for Alzheimer's disease. Attempts have been made to replace some of the missing neurotransmitters that are normally released from brain nerve endings, but these have so far showed only limited benefit. Overall, care for patients with Alzheimer's disease and dementia in general requires support from both the family and community services. It is a most difficult and testing time for the family, especially as long-term institutional care for these patients is still at a premium.

### Organ grinding: loss of other bodily functions

Like the brain, the heart and kidneys have their maximum number of cells at birth. As we grow older the number of cells in both of these organs diminishes. For example, at birth each kidney contains up to one million filtering units – nephrons. These nephrons filter blood and discharge the waste products of our body's metabolism into the urine. If we become thirsty our kidneys are good at conserving water, and if we drink a lot, they are good at getting rid of it. When we are young, our kidneys produce more urine in the day, but as we get older this rhythm is disturbed. More urine is produced at night, often causing great inconvenience. As women age, the sphincter muscles around the urethra (the tube from the bladder to the outside) become progressively weaker, making incontinence a common – and distressing – problem. Older men have a tendency to the opposite problem. An enlarged prostate gland may block the stream of urine through the urethra.

One of the main waste products excreted in urine is known as creatinine, which, along with other waste products, is toxic to the body in high concentrations. If our kidneys slow down, the level of creatinine in our blood increases; if our kidneys stop working altogether then creatinine and excess water need to be removed by dialysis. As we grow older, we gradually lose nephrons; by the age of 80, many people will have lost almost half the number they had at birth, and the kidneys eventually shrink in size. Yet creatinine does not accumulate in older people's blood as might be expected. The explanation for this phenomenon is found in another ageing process happening in parallel with the loss of kidney function: a loss of muscle bulk. Creatinine is generated in muscles and the more muscular we are, the higher our blood creatinine levels. From about the age of 25, an average person will lose muscle and replace it with fat. The ageing kidneys therefore have less creatinine to excrete than those of a younger person. As we grow older we should lose weight to compensate for this loss of muscle bulk, but most of us put it on.

Our old man's heart has been weakened over a long lifetime of working at pressure. The rhythm of his heart is controlled by special clusters of pacemaker cells; by his age, 90 per cent of these irreplaceable cells have died, with inert scar tissue in their place (see Chapter 5) When he was 20 years old, his heart rate could reach nearly 200 beats per minute. But since then, his maximum heart rate has slowed by about one beat per minute every year. His top heart rate is now about

135 beats per minute, which limits the amount of oxygen-rich red blood cells he can get to his muscles. He used to be able to run at a top speed of 10 miles an hour. Now he can just about manage a mile or two.

Over a lifetime and depending to an extent on where we live, our lungs will have inhaled millions of litres of polluted air. Tiny particles of carbon and other pollutants cause inflammation and fibrosis in the lungs. If you live in a busy town, by your seventies your lungs will probably be as stiff as old leather. They'll be less able to exchange carbon dioxide for oxygen than the lungs of someone the same age living in the country. Our elderly man is fortunate in living near the sea, with more chance of breathing fresh air. Even so, by now his lungs have lost 15 per cent of the delicate surface which absorbs oxygen into his blood, and their normally elastic walls have stiffened. It's a good job he doesn't smoke – for of course smoking also has a major influence on lung function and increases the risk of cancer (in many other organs, as well as the lungs).

While our old man has taken steps to preserve his health in some directions, he can do little to (literally) save his skin. As I pointed out on page 152, the skin is particularly vulnerable to cell death; in any case, the skin of old people is lost faster than it can be replaced. At his age, our man loses a staggering one billion cells every day. After the age of 30, our skin gets thinner. This has knock-on effects. The thinner our skin, the less able it is to protect the tissues beneath and, as a result, it loses its natural elasticity and becomes wrinkled. For the same reason that lamb is more tender than mutton, our muscles and tendons also lose their elasticity and become stiff. This happens much more quickly in those who lead sedentary lives. When we walk our muscles are instructed to move by signals from the brain. But in our old man's spine some of the nerve cells that carry these signals have been killed by free radicals. Lacking stimulation, his leg muscles have begun to waste away, in the same way as they would if he hardly ever used them. Every step he takes now is an effort.

In the last few years, other parts of his body have been deteriorating. For 87 years the lens in each eye has been maintaining itself by laying down coats of a transparent protein on its surface. As each layer is put down, the lens has become thicker and thicker; his eye muscles struggle to focus the lens. Cataracts are formed when the lens actually becomes opaque. Since the age of 50, he has been long-sighted. (If he had previously been short-sighted, he might now be juggling two pairs of glasses depending on what he was looking at.)

As we grow older layers of protein make the lens of the eye increasingly thick. This makes it hard work for the tiny muscles focusing the lens, and explains why older people become long sighted.

From the age of 30 we become increasingly deaf. As we grow older, the nerve from the ear becomes desensitised and hard wax blocks the passage of sound waves in the tubes of the outer ear. Initially we lose appreciation of high-pitched sounds, then conversation in a noisy environment becomes difficult. This loss of hearing is common in old age, but that doesn't mean it should be accepted as normal. Hearing aids are becoming increasingly sophisticated and comfortable, while something as simple as syringing wax out of the ears can make a marked improvement to quality of life.

Teeth, however, fall out because of poor dental hygiene rather than age – improved dental care effectively prevents decay. But even if teeth do drop out we can replace them with dentures. A lack of teeth no longer means we have to live on slops or starve to death (a common consequence of tooth decay in wild animals, and probably our ancient ancestors). If Shakespeare were writing today, he wouldn't have to describe the seventh and last stage of man as '*sans* teeth, *sans* eyes, *sans* taste, *sans* everything'.

## A sensible life: diet and exercise

While many a bizarre theory claims to stave off the effects of ageing (see box, page 167), a more rationally motivated approach has looked at the effect of reducing calorie intake. This is known to prolong life in rats (see box, page 152), but there is currently no evidence it works the same way in humans. Certainly, obese people don't live as long as those of normal weight, but that is because they die of conditions such as heart disease, not because they age more quickly.

Lack of exercise has obvious effects such as atrophying muscles and increased weight gain. But, inside the body, blood levels of glucose, cholesterol and other fats go up and put the sedentary couch potato at risk of cardiovascular disease.

The metabolic rate of a cell from a calorie-restricted rat is the same as a normally fed rat. However, dieting rats suffer less cell damage and ageing due to a reduction in the damaging by-products of oxygen metabolism – oxygen free radicals. However, when we exercise (and even for a period of time after we have finished exercising) our metabolic rate increases and so, one would assume, do the number of free radicals we produce. Can exercise really be harmful? It is unclear whether too much exercise can hurt us, but moderate exercise does keep us in good physical shape and prevents the onset of cardiovascular disease. In this respect exercise helps us live longer, but it does not delay the ageing process.

# FINAL CURTAIN

Our old man will die today. At his age, any number of conditions could bring about death; what actually precipitates it is a bleeding stomach ulcer. This has been developing for some time, but he hasn't been aware of it. In any case, diseases in elderly people often reveal themselves in an unusual way. An organ may become newly diseased but, before this causes symptoms, another organ – one made vulnerable by ageing – will be less able to work harder in an emergency: and an emergency awaits our old man.

After a normal routine morning of walking the dog and visiting his family, the old man takes lunch. In his stomach, the food is broken down by digestive acid. Until now, his stomach has protected itself from this acid with a rapidly regenerating cellular lining. But, just like everywhere else in his body, the stem cells which generate this lining have been dying off. His stomach walls are getting threadbare and open to acid attack. In his increasingly fragile state, a condition that poses no real danger to most people starts to become life-threatening. In his stomach, digestive acid has burnt its way right through an ulcer into an artery. Blood pours through the hole and out into his stomach. He feels queasy and light-headed. He makes his way to bed, and lies down.

Our old man's body is in crisis. If he were younger, his perforated artery could instantly constrict to slow the loss of blood. As it is, his 87-year-old arteries have furred up with atheroma and become hardened with calcium; they cannot constrict and so bleeding continues unabated. Many of the muscle cells in his heart

Keeping physically active is another way to delay the ageing process. As we age, our muscles and tendons stiffen up, our heart beats less strongly and nerve impulses travel more slowly. Keeping active improves the quality of life as we grow old.

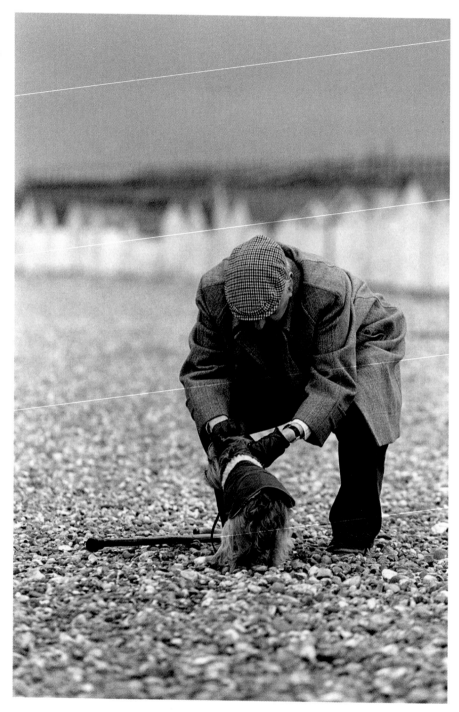

have stopped working so the heart doesn't beat quickly enough or strongly enough to compensate for the loss of blood.

Blood continues to pour into his stomach. His blood pressure is dropping, so his heart beats faster to keep blood moving around his body. But, even pumping at its maximum rate, his fragile heart cannot restore his blood pressure to normal. This is the beginning of the end.

The drop in blood pressure means that less blood and oxygen pass to the brain. For all its power, the brain is a fragile organ; and our man's is already vulnerable from the loss of millions of cells over many decades. The slightest period of oxygen deprivation is potentially lethal to the cells of the brain. Now they are getting only 75 per cent of their usual oxygen supply. To protect itself from oxygen starvation, his brain suspends all activity except that which controls his basic body functions, like breathing. He is slipping into unconsciousness.

Blood from the stomach ulcer passes into the bowels where it is partially digested. The digested red cells release potassium into the bloodstream. High concentrations of potassium in the bloodstream can be lethal even to a young person. It is normally eliminated by the kidneys, but as the kidneys of our old man have lost a great deal of their function, potassium levels in the blood accumulate – high levels have their most toxic effects on the heart. The heartbeat becomes irregular, making it even harder to pump effectively. For the first time in nearly 90 years, his heartbeat falters. His heart is pumping so little blood, it can no longer support its own beat. This vital organ, which has sustained his body through every moment of his life, is grinding to a halt.

**Left**: An artery at the base of a stomach ulcer spurts blood. Old people are less able to adapt to diseases than the young. A minor physical problem in a young person can be a disaster in the elderly.

**Overleaf**: As we grow old and forget skills learnt long ago, the connections between brain cells start to pull away. It is the loss of these connections that may be more important to our mental decline than the loss of individual brain cells.

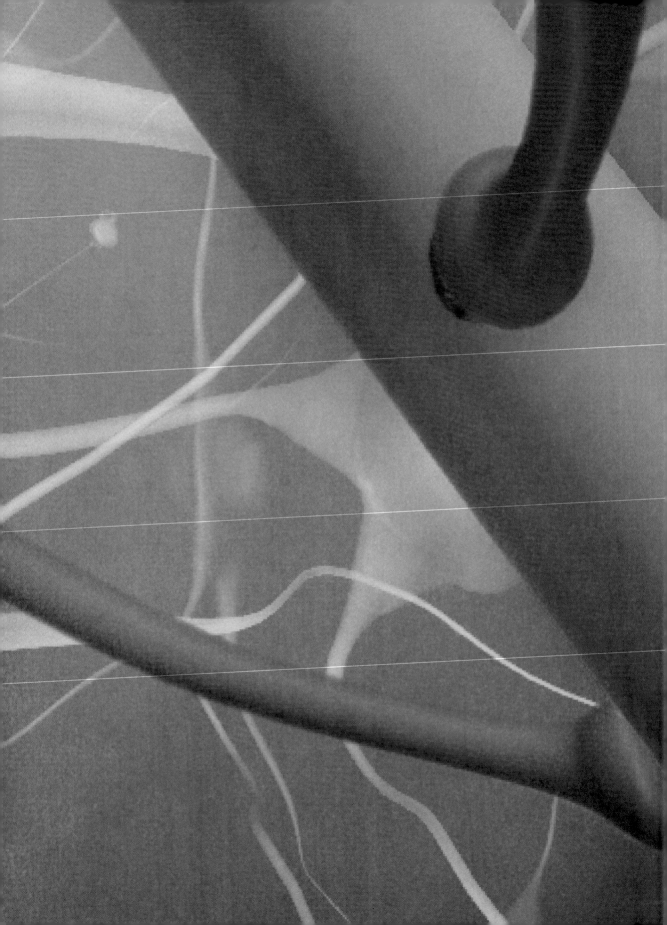

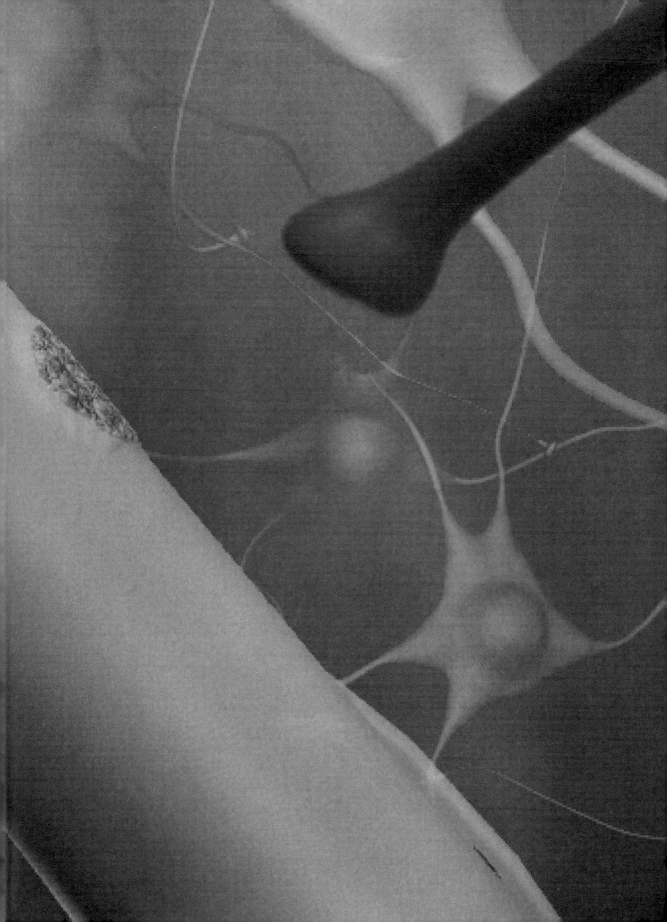

Death is the only certainty in life. This man's genes continue through his son and grandson.

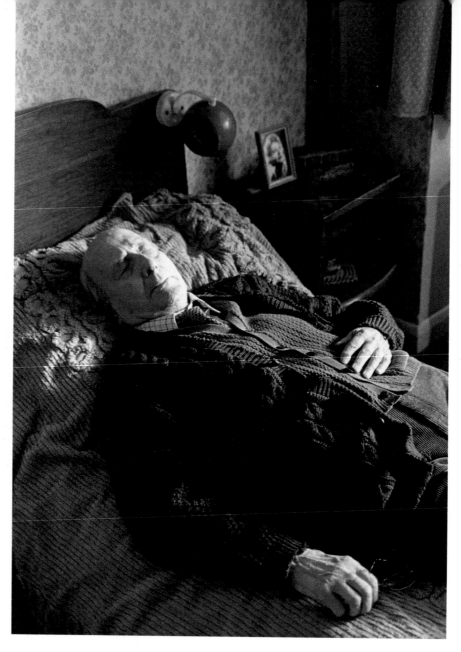

The traffic around his body has ceased. He is relaxed and tranquil; he feels no pain because, starved of oxygen, his brain cells release waves of morphine-like chemicals called endorphins. They have an anaesthetic effect on his body. As the brain cells in his visual cortex begin to die, they fire off signals. Although his eyes are shut, it is as if he's seeing. This may explain the tunnel of light which many people have reported on being brought back from the brink of death.

When brain cells die they detach from one another, breaking the connections

## Living for ever? Elixirs of life

There has always been a huge market for anything that promises to prolong life: pills and potions, foods and fads. Most of them have absolutely no scientific rationale, let alone evidence that they work. At present, one can make suggestions about a healthy way of living and so prevent the onset of disease, but no solid claims can be made for any diet or lifestyle that slows ageing in humans. For thousands of years, ginseng (a plant root) has been given as a mild stimulant to overcome tiredness, but there is no evidence that it prolongs life. Outrageous claims have been made for hormone implants, plant extracts, honey or even the local anaesthetic lignocaine; none has been shown to prolong life.

If oxygen free radicals (oxidants) are damaging to cells (see box, page 152), it is not surprising that great claims have been made for anti-oxidants as protectors of cells. Large quantities of the anti-oxidant vitamins A, C and E have been taken in the hope that they must do some good. Having won two Nobel Prizes, the scientist and pacifist Linus Pauling (1901–94) laid great claim to the benefits of vitamin C. He thought that huge doses of this vitamin could enhance and prolong life, prevent cancer and improve resistance to infection. For vitamin C to have these properties he really did mean huge doses. For example, the US government recommend a daily intake of 60 milligrams of vitamin C, while Pauling advocated doses of 3,000 to 12,000 milligrams. At present there is no evidence that the effects of huge doses of vitamin C slow ageing, but there is some controversial evidence that it may reduce the incidence of some cancers. Several other clinical trials are still under way. Even if they don't do much good, such huge doses of vitamin C in most people probably don't do much harm either. However, there are some people at risk of kidney stones who will become even more vulnerable if they take such vast quantities.

There is also no role for vitamins A and E as agents to slow down ageing. Indeed, vitamin A given to a large group of heavy smokers has been shown to increase the likelihood of lung cancer. A sensible balanced diet with lots of fruit and vegetables, low in fats and not too many calories, appears – as ever – the most sensible advice to prevent premature death from disease, rather than slow down ageing. Not that this will discourage new generations of hucksters and charlatans from exploiting a gullible public's emotions and pockets.

they made during a whole life of thinking and remembering. As his last brain cell expires, his most basic body function ceases.

Death is a process in which every tissue and organ proceeds at its own pace. New skin cells will continue to grow for another 48 hours. But eventually this too will stop. Many people believe that nails and hair still grow after death, but these are myths. As cell after cell ceases its activity, his body stops generating heat of its own. With its internal fire extinguished, his body starts to cool down towards the temperature of the room.

Although our old man is dead, he still has a biological presence. The blueprint of his cells, his DNA, has been passed down to his son, and from his son to his grandson. Though his consciousness is gone, his characteristics may live on. In time his son will grow old, and his son in turn: the cycle of life and death that will last as long as the human race.

# Epilogue

These six biological dramas have graphically illustrated common life events, and how we respond both physically and mentally. From the minute we are born until the day we die, life is like stepping through a medical minefield. In the industrialised nations, the field becomes more densely packed with mines the older we become. In the developing world, childhood is probably the most dangerous stage – poverty, malnutrition and poor sanitation take a heavy toll, just as they did in the UK 150 years ago. At that time, half the children born would not see their tenth birthday. Major changes in public-health policy established clean water and sewerage systems. The consequent impact on average life expectancy was much greater than any medical advance. In fact, our own bodies may be our greatest medical asset. Our remarkable capacity for self-healing and adaptation to injury or disease have kept us going through the harshest conditions and most severe illness.

Even in countries that regularly boast new medical advances, the impact on the quality and quantity of human life has been limited. Trumpeting medical 'success stories' has, however, increased public expectation of medical ability and to some extent generated a *laissez-faire* attitude towards physical health. There is no room for complacency. 100 years ago, in the UK, an average 65-year-old man could expect to live another 9.5 years; these days, after all our developments, that expectation has increased by only a few years, to 13.2 years. Furthermore, contemporary living has created its own new threats to health. For example, a high-fat diet and sedentary lifestyle make us prone to heart disease, while motor vehicles have exposed us to serious accidents and the recent explosion of addictive drugs remains an often fatal temptation. For all its complexity and resilience, the human body remains vulnerable.

As we have seen, naturally occurring infections are still a threat to us, particularly those that can mutate rapidly, such as the flu virus. A global pandemic could prove fatal to millions – and imagine the virulence of HIV if it could be transmitted by a sneeze! If natural threats were not enough, we seem perfectly able and willing to manufacture our own – in the shape of weapons of mass destruction and reckless abuse of the environment. In dominating the world, we should not think ourselves so separate from other animals - after all, we share about 99 per cent of our genes with the chimpanzee. Nor should we take our survival for granted: modern man has lived for barely 100,000 years, a blink of an eye compared to the 165-million-year dominance of the dinosaurs. Yet often a kind of arrogance seems to blind us to the true facts of life. We are living close to the edge.

Of course, human ingenuity has brought about significant advances in medical science that have benefited many, by helping to defuse some of life's medical minefields. In this case at least we have come an incredibly long way in a very short time. A lifetime ago, H.G. Wells saw human history as 'a race between education and catastrophe'. Perhaps, if the body's story is rewritten a lifetime from now, we may still be in the running.

# GLOSSARY

**alcoholic unit**: a method of measuring alcohol consumption. One unit = one glass of table wine or half a pint of beer or one measure of spirits or one measure of sherry. Standard bottle of spirits = 32 units; standard bottle of wine = eight units; can of extra strong lager = four units. Recommended maximum alcohol intake for men: 3–4 units per day; for women: 2–3 units per day.

**antibody**: a protein molecule (of the class immunoglobulin) that is produced by cells of the immune system (B cells), specifically targeted against an invading organism or particle (an antigen).

**anti-oxidants**: a substance (such as vitamin C and E) that counteracts free radicals – the potentially damaging products of normal cell metabolism.

**atherosclerosis**: degeneration of the inner layers of arteries, formed by the build-up of atheroma.

**atheroma**: a mass or plaque of cholesterol and other fatty material on the inner layer of arteries. Rupture of a plaque in a coronary artery can lead to a heart attack (myocardial infarction).

**bacteria**: a single-celled micro-organism, about one micrometre in diameter, which is capable of causing disease. There are thousands of different types of bacteria, sensitive to different types of antibiotics.

**biological clock**: an innate body clock, controlled by the pineal gland deep within the brain. It controls many physiological activities including our sleep-wake pattern and the release of some hormones.

**body-mass index (BMI)**: a measure of height for weight. Calculated as weight in kilogrammes divided by height in metres, squared. The average range is between 19 and 26.

**brain death**: irreversible brain damage causing the end of independent breathing. Diagnosed by specific tests carried out by two doctors on two separate occasions. It is indicative of death.

**calcium**: an element whose ions and salts are essential to many biological functions. 99 per cent of the body's calcium is stored in bone.

**cartilage**: a firm but flexible tissue that lines many joints and forms the bulk of the infant skeleton, until it is replaced with bone.

**CAT scan**: computerised axial tomography scan, which uses several beams of X-ray orientated in different directions, then analysed by a computer to create an image of the body (most often the brain) in cross-section.

**cell**: unit of organic function from which the whole body is made; approximately 100 trillion in the human body.

**cerebral cortex**: the grey matter (made up of billions of brain cells) that exists on the outer surface of the brain. In evolutionary terms, it is the most recently developed part of the human brain: it is also the largest. It can be divided into about 50 different areas that control different bodily functions.

**cholesterol**: a member of the steroid group of molecules, which is found in many different tissues of the body. Most cholesterol is made in the liver, but some is absorbed from the diet. High levels of cholesterol in the blood can lead to atherosclerosis and heart disease.

**chromosome**: a rod-like structure of DNA and protein found in the nucleus of a cell and made up of genes. There are normally 46 chromosomes (23 pairs) in a body cell, including the two sex chromosomes (XY = male or XX = female).

**cloning**: a process of asexual reproduction (natural or artificial) by which a new organism is produced with the identical genetic material as its parent. Sheep have now been cloned by replacing the nucleus of an unfertilised ovum (egg cell) with that of a nucleus from an ordinary body cell and putting the altered ovum into the womb.

**collagen**: a protein substance that puts much of the strength into connective tissue, such as cartilage, skin, bone and tendons.

**dementia**: a disorder of the brain marked by deteriorating memory, impaired reasoning and personality change. It is associated with a progressive loss of brain cells. The most common type of dementia is Alzheimer's disease but atherosclerosis, causing the arteries to the brain to be narrowed, can also be responsible.

**DNA**: deoxyribonucleic acid, a self-replicating material from which genes are made. The exact composition of DNA in a gene determines the composition of a protein to be made and consequently the function of that particular gene.

**dopamine**: a neurotransmittter, a molecule released from the ends of brain cells. It plays a role in the brain circuits responsible for pleasure and reward and also in the part of the brain which co-ordinates body movements. Disease in this area can lead to Parkinson's disease, characterised by slow, awkward movements. It is treated by replacing dopamine.

**endometrium**: the inner lining of the womb, which is shed each month during a menstrual cycle. If an egg is fertilised, it will embed in the endometrium, which is consequently conserved for the duration of the pregnancy.

**endorphins**: naturally occurring brain substances that bind to opioid (a word derived from opium) receptors in the brain and so raise the threshold for pain.

**enzyme**: a naturally occurring catalyst that speeds up the rate of a biochemical reaction.

**free radicals**: unstable molecules that contain unpaired electrons and react damagingly with other molecules. Normal oxygen-fuelled

metabolism within a cell is responsible for producing free radicals. The long-term consequences of free radical damage to a cell are probably the main cause of ageing.

**growth plates**: the ends of long bones containing specialised cells that pump out new bone until adult dimensions are reached.

**hippocampus**: an area of the brain which channels our experiences into the parts of the brain that store memory. The nerve connections to and from the hippocampus become progressively destroyed in people with dementia.

**hormone**: a regulatory substance produced by a gland and then transported in the bloodstream to another part of the body to carry out its function.

**HRT**: hormone replacement therapy. Theoretically this could apply to the replacement of any hormone found in reduced amounts (e.g. thyroid hormone when the thyroid gland is underactive). However, HRT has become synonymous with the replacement of oestrogen to post-menopausal women whose ovaries have stopped producing the hormone. HRT is of value in reducing the incidence of thin bones (osteoporosis), and possibly heart disease, in postmenopausal women.

**hypothalamus**: part of the brain that controls many of the automatic functions of the body. It is also critical to the control of hormones released from the master gland – the pituitary gland.

**ischaemia**: a reduction of blood supply to an organ, due to a blockage (such as atheroma) or physiological constriction of the blood vessel. Ischaemia to heart muscle (due to narrowing of a coronary artery) causes chest pain – angina.

**limbic system**: an area of the primitive (in evolutionary terms) brain that is common to all mammals. It is concerned with smell, emotion and certain automatic functions of the nervous system.

**macrophages**: cells that are part of the immune system. Their main role is to clear up cell debris following infection or inflammation. They are attracted to a damaged area in the body by the chemicals produced by inflamed tissue.

**melatonin**: a hormone produced by the pineal gland in the brain, which influences our biological clock. It causes sleepiness and is produced in high amounts during the middle of the night.

**MRI scan**: magnetic resonance imaging, a type of scan that does not use X-rays. Instead, it maps the distribution of hydrogen ions in the body using radio-frequency energy in a strong magnetic field. It produces very sharp images of our internal organs. Recent advances of MRI are allowing images of brain function.

**oestrogen**: a hormone produced by the ovary that gives women their secondary sexual characteristics. Much less oestrogen is found in the bloodstream following 'shut down' of the ovaries at the time of the menopause. It is replaced in HRT.

**osteoporosis**: thinning of the bones, making them more fragile and susceptible to fracture. More common in women after the menopause and in older people generally.

**pathogens**: organisms that cause disease.

**pheromone**: a chemical substance secreted by an animal to attract a mate. In humans, testosterone is converted into a substance called androstenol that is carried in sweat to the skin surface. Bacteria on the skin feed on androstenol and give off a strong scent – probably the human equivalent to a pheromone.

**plasma**: the fluid portion of blood in which all blood cells are carried.

**platelets**: specialised cell fragments that take part in blood clotting. They do not contain a nucleus.

**progesterone**: a hormone produced by the ovary to prepare the womb for implantation of a fertilised egg. If no such egg implants, then the levels of progesterone in the bloodstream fall and the endometrium is shed (menstruation begins).

**protein**: a molecule made up of amino acids. Each protein is coded by a specific gene.

**RNA**: Ribonucleic Acid. The genetic information that determines the order of amino acids in a protein is encoded by DNA. RNA has a structure exactly complementary to DNA, but can move from the nucleus of a cell to its protein-building machinery.

**stem cells**: the basic cell factories from which many different types of cell develop.

**steroid**: a group of molecules with a characteristic chemical structure. They include many hormones and vitamins, not just anabolic steroids (which build up muscle bulk).

**synapse**: the junction between two nerves, or between a nerve and a muscle. Nerve impulses reach a synapse and then a neurotransmitter such as dopamine is released in order to propagate the message.

**testosterone**: a steroid hormone produced mainly in the testicles and responsible for secondary sexual characteristics in the male. It is also responsible for sex drive in men and women (it is produced in some body tissues other than the testicles).

**virus**: these are the smallest known infective organisms, up to 100 times smaller than bacteria. They are inert outside of a host cell and must therefore invade a cell and take over its protein-making machinery in order to replicate. The fact that viruses invade the body's cells makes it difficult to kill them without damaging the host cell.

# INDEX

achondroplasia 41
acromegaly 41
adrenaline 81, 129, 132
ageing 145–69
aggression 81–3
AIDS 99
Albanian paradox 121–2
alcohol 37, 69, 71–85, 123–5
Alzheimer's disease 159
amantadine 108
amniocentesis 16
anabolic steroids 68
androstenol 73
angina 129
angiotensin converting enzyme inhibitors (ACE-inhibitors) 137
antibiotic resistance 88, 95, 96
antibiotics 88, 96
antibodies 99
anticholinergics 88
antigenic drift 101
antigenic shift 100–1
antihistamines 88
anti-oxidants 167
antiviral drugs 108
aorta 115
aphrodisiacs 68
aspirin 88, 96, 100, 133, 136
atheroma 127–8, 163
atherosclerosis 115, 118
atria 112, 114–15
autonomic nervous system 63, 80

balloon angioplasty 140
B cells 98–9, 100, 106–7
beta-blockers 136–7
biphosphonates 37, 48
birthing pools 23
blood-brain barrier 71

blood clots 48, 49–50, 129–30, 133
blood pressure 20, 21–2, 115, 120, 122–3, 133
body-building 68
body-mass index (BMI) 121
bone mineral density (BMD) 53
bones 33–55
brain 57–85, 156–9, 165, 166–7
brain death 64
brain stem 64
Braxton Hicks contractions 28
breathing 26, 128
brittle-bone disease 48
Broca's area 61
bruises 51

caesarean section 30, 31
calcium 26–7, 34, 37, 38, 39
calcium channel blockers 34
cancer 109, 120, 152–3
cardioplegia 140
cardio-pulmonary resuscitation (CPR) 136, 141
cardiovascular disease 119–20
carpal tunnel syndrome 28
cartilage 38, 42
castration 66
cataracts 161
cerebellum 82, 84
cerebral cortex 58–60, 63, 64, 65, 72, 81
chest infections 114
chicken flu 93, 105
chickenpox 91
cholesterol 118, 125
cilia 92
circadian rhythms 85

cirrhosis 83
cloning 15
cocaine 72, 81
collagen 34, 48
Colles fracture 42
common cold 88, 89
complete heart block 115
computerised axial tomography (CAT) scans 55, 59, 159
contractions 28, 29
coronary angiogram 138–9
coronary arteries 115, 126, 129
coronary artery bypass graft (CABG) 138–40
creatine kinase MB 130
creatinine 160
Creutzfeld Jacob disease (CJD) 41
cystic fibrosis 15, 109

death 169
decongestants 88
defibrillation 136, 138, 139, 140
dementia 159
diet 121–2, 163, 169
DNA 151, 169
dopamine 70–2
Down's syndrome 16, 156
dual X-ray absorptiometry (DEXA) scan 55
dwarfism 41

ectopic pregnancy 13
eggs 12–16
ejaculation 80
electrocardiogram (ECG) 131, 132, 138
embryo 17–22
endorphins 45, 166
epidurals 22
erection 80
evolution of the mind 64–5

exercise 122, 133–4, 153, 163
eyes 161

fenfluramine with phentermine (fen-phen) 121
fertilisation 12, 14–16
fibrin 48
fibrous joint 42
flu virus 88–108
foetal heart 113
foetus 22–9, 91
folic acid 16
food 69–70
forceps delivery 30–1
fractures 42, 43–55
free radicals 151, 152, 153, 163, 169
frontal leucotomy (lobotomy) 61

gene therapy 108–9, 123
genetics 15–16, 89
German measles see rubella
gestational diabetes mellitus 23–4
gigantism 41
goblet cells 92
green stick fracture 42
growth 38–41, 66
    disorders 41
growth cartilage 38
growth hormone 39, 41

haemagglutinin spikes 92, 103
haematoma 49–50, 53
haemoglobin 51
haemolytic disease 17
hearing 63, 162
heart 111–43, 160–1, 165
    attacks 115, 126–43
    disease 119–20, 149
    transplants 141–3

height 40, 66

heredity 123

herpes virus 91

high density lipoprotein (HDL) 122, 125

hippocampus 159

histamine 95

HIV 99, 105

home births 23

hormone replacement therapy (HRT) 12, 37, 67, 119–20

human chorionic gonadotrophin (HCG) 18, 23

human placental lactogen (HPL) 23

hypothalamus 65, 68, 70, 80, 85

hypothermia 156

immune system 21, 79, 85, 87–109, 141, 149

infertility 14, 68

influenza virus 88–108

insulin 23–4

intermittent claudication 127

intracytoplasmic sperm injection (ICSI) 14

in vitro fertilisation 14

ischaemia 115, 138

jaundice 83

joints 42–3

kidneys 26, 40, 83–4, 160, 165

labour 22, 29–31

language 60

life expectancy 37–8, 146–9

ligaments 42

limbic system 63, 70–2, 78, 80–1

liver 83, 85

low density lipoprotein (LDL) 125

lungs 25–6, 100, 112, 114, 133, 134, 161

lymph glands 97–9

macrophages 94, 95, 96, 97, 99

magnetic resonance imaging (MRI) scans 55, 59

male menopause 67

median nerve 27–8

melatonin 85, 149

memory 159

memory cells 100

menopause 12, 67, 119

menstrual cycle 12–14

muscles 161
  joints 42
  nerves 64
  pregnancy 28–9

mutation 101, 108

myalgic encephalitis (ME) 89

myocardial infarction 115, 130

myocardium 115

nerve impulses 64

neuraminidase 92, 103

oestrogen 23, 25, 67, 83, 119, 120, 149

orthopnea 134

ossification 38

osteoarthritis 42

osteoblasts 35–6, 50, 55

osteoclasts 35, 37, 55

osteocytes 35, 36

osteogenesis imperfecta 48

osteomalacia 39, 40

osteoporosis 37–8, 39

ovulation 12

oxytocin 27, 29

pain 45, 48, 55, 84
  angina 129
  atheroma 127–8
  labour 22, 29–30
  virus response 96

paralysis 34

pheromones 69, 73, 77

phrenology 59

pituitary gland 41, 65, 68

placenta 17, 18, 19–21, 26

platelets 48, 115, 136

pneumonia 100

pre-eclampsia 18, 21

pregnancy 13, 16–30, 91

progeria 156

progesterone 22, 23, 67

pulmonary oedema 133

pulmonary tuberculosis 114

pulmonary veins 114

rate of living theory 152, 156

red blood cells 21, 112, 116–17

retirement 149

Reye's syndrome 100

rhesus blood groups 17

rickets 39, 40

rubella 91

serotonin 25

sex drive 65–9

sexual arousal 72–8

shingles 91

skin 48, 51, 66, 152, 161

skull 42

slipped disc 42, 64

smallpox 98, 102, 104

smell 77–8

smoking 37, 72, 92, 122, 124, 161

Spanish flu 103

sperm 12, 14–16

spina bifida 16

spinal cord 42, 64, 84

stem cells 150

Streptokinase 133, 134

sweat 73

swine influenza 103, 105

synapses 64

Syndrome E 81

T cells 98–9, 100, 104, 106–7

teeth 162

temperature
  ageing effects 155–6
  death 169
  during pregnancy 22–3
  virus response 96

tendons 42

testosterone 65–9, 73, 78, 81–3

test-tube babies 14

tetanus 51

tetany 34, 40

touch 79–80

troponin 130

vaccination 51, 91, 94, 98

valves (heart) 112

ventouse extraction 30–1

ventricles 112, 114–15

ventricular fibrillation (VF) 136, 137

vertebrae 42, 64

viruses 88–109

vitamins
  A 167
  C 88, 167
  D 26, 37, 39–40
  E 167

weight 20, 24, 119, 120–1, 163

white blood cells 95, 115

xenotransplantation 141–3

X-rays 55, 59

pacemaker 115, 127–8

# BIBLIOGRAPHY

Throughout this book I have used references from many medical journals, but especially the *Lancet, British Medical Journal, New England Journal of Medicine* and *British Journal of Obstetrics and Gynaecology.*

Alberts, B., D. Bray, J. Lewis, M. Roff, K. Roberts, J. D. Watson, *Molecular Biology of the Cell*, (Garland Publishing Inc., 1989).

Austad, Steven N., *Why We Age*, (John Wiley and Sons, Inc., 1997).

Beatty, Jackson, *Principles of Behavioural Neuroscience*, (Brown and Benchmark, 1995).

Bennett, G. and Ebrahim, S., *The Essentials of Healthcare in Old Age*, (Edward Arnold, 1995).

Blackmore, S., *Dying to Live*, (Prometheus Books, 1993).

Collier, Leslie and John Oxford, *Human Virology*, (Oxford University Press, 1993).

Coni, N., W. Davison and S. Webster, *Ageing: The Facts*, (Oxford University Press, 1992).

Echert, R., D. Randall and G. Augustine, *Animal Physiology: Mechanisms and Adaptations*, (W. H. Freeman and Co., 1988).

Garrett, Laurie, *The Coming Plague*, (Penguin Books, 1994).

Greenfield, Susan, *The Human Brain (A Guided Tour)*, (Weidenfeld and Nicolson, 1997).

Harper, David R., *Molecular Virology*, 2nd edition, (Bios Scientific Publishers, 1998).

Hytten, Frank and Geoffrey Chamberlain (eds.), *Clinical Physiology in Obstetrics*, (Blackwell Science, 1991).

Levy, Jay, Heinz Fraenkel-Conrat and Robert Owens, *Virology*, 3rd edition, (Prentice Hall, 1994).

Newland, S. B., *How We Die*, (Chatto and Windus, 1994).

Newland, S. B., *The Wisdom of the Body*, (Chatto and Windus, 1994).

Porter, Roy, *The Greatest Benefit to Mankind: A Medical History of Humanity from Antiquity to the Present*, (HarperCollins, 1997).

Rose, Steven (ed.), *From Brains to Consciousness? Essays on the New Sciences of the Mind*, (Allen Lane, The Penguin Press, 1998).

Weatherall, David, *Science and the Quiet Art*, (Oxford University Press, 1995).

Williams, P. (ed.), *Gray's Anatomy*, 38th edition, (Churchill Livingstone, 1995).

*Oxford Textbook of Medicine*, 3rd edition, (Oxford University Press, 1996).